WITHDRAWN

The Flow of High Polymers

Continuum and Molecular Rheology

STANLEY MIDDLEMAN

Chemical Engineering Department
University of Rochester
Rochester, New York

1968
INTERSCIENCE PUBLISHERS

A DIVISION OF JOHN WILEY & SONS, New York · Chichester · Brisbane · Toronto

Copyright © 1968 by John Wiley & Sons, Inc.

All rights reserved.

Reproduction or translation of any part of this work beyond that permitted by Sections 107 or 108 of the 1976 United States Copyright Act without the permission of the copyright owner is unlawful. Requests for permission or further information should be addressed to the Permissions Department, John Wiley & Sons, Inc.

Library of Congress Catalog Card Number: 67-29460
Printed in the United States of America

10 9 8 7 6 5 4 3 2

To my wife, Jo Ann,
and to my daughters, Melissa and Sharon

Preface

This book is written for people interested in the flow of real materials, particularly those polymeric materials which have played such an essential role in the growth of chemical industry in recent years. Three major topics are discussed: the *definition* and *measurement* of material properties, the *prediction* of these properties from theoretical principles, and *correlation* techniques with which these properties may be predicted from related experimental data.

Chapter 1 gives a brief review of physical principles which underlie the subsequent analyses. It is assumed that the reader has a background in fluid mechanics. Such a background is required in Chapter 2, but the reading of subsequent chapters does not really depend on any ability to derive the relationships presented in Chapter 2.

In Chapter 2 analyses are given of flows commonly used in the study of polymeric fluids. These analyses generally require assumptions which simplify the dynamic equations to the point that analytical solutions are possible. Where possible an attempt is made to discuss the treatment of data which are subject to perturbations arising from the failure of simplifying assumptions to hold exactly throughout the experimental system. No attempt is made to present details of equipment design and operation. Sufficient references are made to experimental studies which illustrate the use of the analyses presented that this aspect of experimental detail may be found if desired.

Recent years have witnessed significant advances in the theory of classification of material response. In Chapter 3 a review of the continuum mechanical formulation of rheological theories is presented. Much of this section requires a background in tensor analysis if the reader is to follow the developments. In the Appendix is a self-contained treatment of the fundamentals of tensor analysis, the depth of which should be sufficient for an understanding of Chapter 3.

After a brief review of Noll's formulation of the simple fluid concept, Chapter 3 gives a review and critique of specific models put forth to describe purely viscous fluids, and viscoelastic fluids. The relationship of nonlinear viscoelastic theories, both as rate equations and integral equations, to classical linear viscoelasticity is pointed out, and a short summary of linear viscoelasticity is included.

In Chapter 4 the molecular nature of polymers is acknowledged, and rheological theories based upon the physics of macromolecules are described. The detailed mathematical development of molecular theories is not presented, but an attempt is made to describe the physical basis of each theory, and the manner in which it differs from other theories. Predictions resulting from various molecular theories are contrasted.

Chapter 5 reviews and illustrates the success and deficiency of various proposals for the correlation of material functions measured over wide ranges of parameters. In this area there is still a considerable need for experimental data obtained with polymers of well-defined structure, particularly with regard to molecular weight distribution. In the case of polymer solutions there is need for careful studies of the concentration dependence of material functions, particularly into the high concentration region of plasticized and oil-extended polymers.

The vigor with which experimental rheology was being pursued during the period in which this manuscript was under preparation will certainly lead to clarification of many topics which were incompletely discussed in the text. If the reader has been prepared to follow this new literature with understanding, and with a feeling of perspective for what has gone before, this monograph will have served its purpose.

STANLEY MIDDLEMAN

March, 1968

Contents

1. **Introduction** 1
   - I. The Science of Rheology 1
   - II. The Dynamic Equations 3
   - III. Response of a Fluid 6
   - IV. The Simple Shear Flow 8
   - V. Simple Elongation 10
   - References 12

2. **Experimental Techniques** 13
   - I. Steady State Flows 13
     - A. Viscosity 13
     - B. Normal Stress Measurement 40
     - C. Elongational Flows 60
   - II. Dynamic Measurements 65
     - A. Oscillatory Flow 67
     - B. Creep and Relaxation 71
     - C. Wave Propagation on a Jet 76
   - References 82

3. **Constitutive Equations—Continuum Theories** 84
   - I. General Considerations of Constitutive Theories . . 85
     - A. Kinematics of Convected Coordinates 85
     - B. The Simple Fluid 91
   - II. Special Classes of Constitutive Equations 99
     - A. The Purely Viscous Fluid 99
     - B. The Viscoelastic Fluid 102
   - References 130

4. **Molecular Theories of Polymer Rheology** 132
   - I. The Polymer Molecule 132
     - A. Spatial Configuration of a Polymer Chain . . . 134
     - B. Elasticity of a Polymer Chain 137

CONTENTS

- II. Viscosity in Dilute Solutions 139
  - A. Viscosity of a Suspension of Spheres 139
  - B. The Free Draining Polymer 140
  - C. Hydrodynamic Shielding 142
  - D. The Theta-Solvent 145
  - E. Effect of Polydispersity 146
- III. Non-Newtonian Viscosity 147
  - A. The Bueche Theory of Viscosity 147
  - B. Extension to Polydisperse Polymers 150
  - C. Molecular Entanglement Theory 156
  - D. The Molecular Theory of Williams 158
- IV. Dynamic Viscosity 160
  - A. Zimm's Theory 160
  - B. Tschoegl's Extension 162
  - C. Relationship to the Maxwell Model 164
- References 166

5. Correlation and Interrelation of Material Functions 168

- I. Viscosity at Low Shear Rate 168
  - A. Dependence on Temperature 175
- II. Non-Newtonian Viscosity 178
- III. Normal Stress Coefficients 184
  - A. Behavior in the Limit of Zero Shear Rate . . . 184
  - B. Dependence on Shear Rate 186
- IV. Relationship of Steady Shear to Dynamic Properties . . 187
- V. Stress Relaxation 192
- References 198

Appendix A. Tensors 200

- I. Cartesian Tensors 200
  - A. Vectors 200
  - B. Cartesian Tensors 205
- II. Curvilinear Coordinates 213
  - A. General Coordinate Transformations 215
  - B. Contravariant and Covariant Tensors 217
  - C. Christoffel Symbols 220
  - D. Covariant Differentiation 223
  - E. Physical Components 226

Appendix B.   Kinematics **230**
    I. Material vs. Spatial Coordinates 230
    II. Deformation   232
    III. Strain   234

Appendix C.   Monomer Structural Units of Some Common Polymers **238**

Author Index **241**

Subject Index **245**

1

Introduction

I. THE SCIENCE OF RHEOLOGY

The study of polymeric fluids is but a small part of the broader field called rheology. Literally, rheology means the study of flow, but the science of rheology has become much broader than this. It has come to include almost every aspect of the study of the deformation of matter under the influence of imposed stress: it is the study of the internal response of materials to forces.

Between the extremes of simplicity of the newtonian fluid and the hookean solid lie materials of great interest. It is fair to say that commercial interest in synthetic polymeric materials has given the greatest impetus to the rapid growth of the science of rheology. One should not discount, however, the advances which have resulted from the intellectual challenge of producing realistic mathematical descriptions of the response of complex materials.

Several broad, qualitative categories of response can be described. If a small stress is suddenly exerted on a solid, a deformation will begin to occur. The material will continue to deform until molecular (internal) stresses are established which just balance the external stresses. The word "deformation" usually means the equilibrium deformation which is established when the internal and external stresses are in balance. Most solids exhibit some degree of elastic response, in which there is complete recovery of deformation upon removal of the deforming stresses. The simplest such body is the hookean elastic solid, for which the deformation is directly proportional to the applied stress. But elastic response may also be exhibited by non-hookean materials, for which the deformation is not linearly related to the applied stress.

Not all materials reach an equilibrium deformation. In a fluid, if an external stress is exerted, deformation occurs, and continues to occur indefinitely until the stress is removed. A fluid response is one in which there is no resistance to deformation. But internal frictional forces retard the *rate* of deformation, and an equilibrium can be established in which the rate of deformation is constant and related to the properties of the fluid. The simplest such fluid is the newtonian fluid, in which the rate of deformation is directly proportional to the applied stress. However, many fluids exhibit a non-linear response to stress and are called, collectively, non-newtonian fluids. Most synthetic polymer solutions and melts exhibit some degree of non-newtonian behavior.

Between the extremes of elastic and fluid response lies a spectrum of combinations of these basic types of material behavior. For example, there is plastic response, wherein a material deforms like an elastic solid as long as the applied stress is below some limit, called the yield stress. If the applied stress exceeds the yield stress, the material behaves as a fluid. House paint is an example of such a material. Brushing imposes stresses sufficiently large that paint behaves like a fluid. When paint lies on a vertical surface in a thin film, however, the stresses which arise from the weight of the fluid are below the yield stress, and the paint remains on the surface to dry as a uniform film.

Another important class of material is the viscoelastic fluid, which resists deformation but at the same time resists a time rate of change of deformation. Thus it exhibits some combination of elastic and fluid response.

In discussing categories of response it is important to note that, in a given experiment, response may depend not only on the material, but also on the time scale of the experiment. Thus, water behaves like a newtonian fluid in ordinary experiments, but, if it is subjected to ultra-high frequency vibrations, it will propagate waves as if it were a solid. The reason for this apparent change in the type of behavior lies in the fact that response is ultimately molecular in nature, and involves the stretching of intermolecular bonds and the motion of molecules past one another. In general, bonds can be stretched very quickly by an imposed stress, since little motion is involved. On the other hand, considerably more time is involved in causing molecules to "flow." Thus, in a stress field with a very short time scale (high-frequency vibration is an example) the stress may reverse itself before molecules have time to move appreciably, and only mechanisms giving rise to elasticity may have time to be excited. Because behavior can depend upon the type of stress field imposed, it is imperative that when one reports the rheological properties of a material he also reports the range of conditions over which these properties were measured. Only in this way can rheological properties be used with any assurance that they have pertinence to the application at hand.

The range of rheological behavior exhibited by known materials is large; new types of behavior await discovery in the large numbers of new materials created by modern technology. The task of rheology is to create some order in the technique and theory of classification of materials. This

INTRODUCTION

monograph considers but a small segment of the field of rheology. It seeks to describe the methods whereby one measures and predicts the properties of materials that flow. Its major concern is with the non-newtonian viscoelastic fluid.

II. THE DYNAMIC EQUATIONS

The response of any element of a body to the forces acting upon that element must satisfy the principle of conservation of momentum. In a continuum this principle is embodied in the "dynamic equations," written in cartesian coordinates as*

$$\rho\left(\frac{\partial v_i}{\partial t} + v_j \frac{\partial v_i}{\partial x_j}\right) = \rho f_i + \frac{\partial T_{ij}}{\partial x_j} \tag{1.1}$$

for an incompressible fluid of density ρ subject to an external force field $\mathbf{f}$. The stress tensor $\mathbf{T}$ is usually split into two parts; the mean normal stress $p = -\frac{1}{3}T_{ii}$, and the excess over the mean due to dynamic stresses $\boldsymbol{\tau}$. For a fluid at rest, $\boldsymbol{\tau} = 0$, and the mean normal stress is just the hydrodynamic pressure. Thus, with† $\mathbf{T} = -p\boldsymbol{\delta} + \boldsymbol{\tau}$, the dynamic equations become

$$\rho\left(\frac{\partial v_i}{\partial t} + v_j \frac{\partial v_i}{\partial x_j}\right) = \rho f_i - \frac{\partial p}{\partial x_i} + \frac{\partial \tau_{ij}}{\partial x_j} \tag{1.2}$$

Of course, Eq. 1.2 represents three equations, one for each direction of the coordinate system.

In addition to the dynamic equations, one has an equation that expresses the principle of conservation of mass, the so-called continuity equation. For an incompressible material this equation is written in cartesian coordinates as

$$\frac{\partial v_i}{\partial x_i} = 0 \tag{1.3}$$

Often one wishes to use cylindrical or spherical coordinates in the solution of a problem. In that case Eqs. 1.2 and 1.3 must be transformed. Tables 1.1, 1.2, and 1.3 give Eqs. 1.2 and 1.3 in cartesian, cylindrical, and spherical coordinate systems.

* Repeated subscripts imply summation over that subscript.
† $\boldsymbol{\delta}$ is the Kronecker delta, or unit tensor. The negative sign in front of p corresponds to the convention that a pressure is a negative stress and a tension is a positive stress. By these definitions of p and $\boldsymbol{\tau}$, $\tau_{ii} = 0$.

Table 1.1
Cartesian Coordinates (x, y, z)

The Dynamic Equations:

x-Component $\quad \rho\left(\dfrac{\partial v_x}{\partial t} + v_x \dfrac{\partial v_x}{\partial x} + v_y \dfrac{\partial v_x}{\partial y} + v_z \dfrac{\partial v_x}{\partial z}\right) = -\dfrac{\partial p}{\partial x}$

$\qquad\qquad\qquad\qquad + \left(\dfrac{\partial \tau_{xx}}{\partial x} + \dfrac{\partial \tau_{yx}}{\partial y} + \dfrac{\partial \tau_{zx}}{\partial z}\right) + \rho g_x \quad (A)$

y-Component $\quad \rho\left(\dfrac{\partial v_y}{\partial t} + v_x \dfrac{\partial v_y}{\partial x} + v_y \dfrac{\partial v_y}{\partial y} + v_z \dfrac{\partial v_y}{\partial z}\right) = -\dfrac{\partial p}{\partial y}$

$\qquad\qquad\qquad\qquad + \left(\dfrac{\partial \tau_{xy}}{\partial x} + \dfrac{\partial \tau_{yy}}{\partial y} + \dfrac{\partial \tau_{zy}}{\partial z}\right) + \rho g_y \quad (B)$

z-Component $\quad \rho\left(\dfrac{\partial v_z}{\partial t} + v_x \dfrac{\partial v_z}{\partial x} + v_y \dfrac{\partial v_z}{\partial y} + v_z \dfrac{\partial v_z}{\partial z}\right) = -\dfrac{\partial p}{\partial z}$

$\qquad\qquad\qquad\qquad + \left(\dfrac{\partial \tau_{xz}}{\partial x} + \dfrac{\partial \tau_{yz}}{\partial y} + \dfrac{\partial \tau_{zz}}{\partial z}\right) + \rho g_z \quad (C)$

The Continuity Equation:

$$\dfrac{\partial v_x}{\partial x} + \dfrac{\partial v_y}{\partial y} + \dfrac{\partial v_z}{\partial z} = 0$$

Table 1.2
Cylindrical Coordinates (r, θ, z)

The Dynamic Equations:

r-Component $\quad \rho\left(\dfrac{\partial v_r}{\partial t} + v_r \dfrac{\partial v_r}{\partial r} + \dfrac{v_\theta}{r} \dfrac{\partial v_r}{\partial \theta} - \dfrac{v_\theta^2}{r} + v_z \dfrac{\partial v_r}{\partial z}\right) = -\dfrac{\partial p}{\partial r}$

$\qquad\qquad\qquad\qquad + \left(\dfrac{1}{r}\dfrac{\partial}{\partial r}(r\tau_{rr}) + \dfrac{1}{r}\dfrac{\partial \tau_{r\theta}}{\partial \theta} - \dfrac{\tau_{\theta\theta}}{r} + \dfrac{\partial \tau_{rz}}{\partial z}\right) + \rho g_r \quad (A)$

θ-Component $\quad \rho\left(\dfrac{\partial v_\theta}{\partial t} + v_r \dfrac{\partial v_\theta}{\partial r} + \dfrac{v_\theta}{r} \dfrac{\partial \theta}{\partial v_\theta} + \dfrac{v_r v_\theta}{r} + v_z \dfrac{\partial v_\theta}{\partial z}\right) = -\dfrac{1}{r}\dfrac{\partial p}{\partial \theta}$

$\qquad\qquad\qquad\qquad + \left(\dfrac{1}{r^2}\dfrac{\partial}{\partial r}(r^2 \tau_{r\theta}) + \dfrac{1}{r}\dfrac{\partial \tau_{\theta\theta}}{\partial \theta} + \dfrac{\partial \tau_{\theta z}}{\partial z}\right) + \rho g_\theta \quad (B)$

z-Component $\quad \rho\left(\dfrac{\partial v_z}{\partial t} + v_r \dfrac{\partial v_z}{\partial r} + \dfrac{v_\theta}{r} \dfrac{\partial v_z}{\partial \theta} + v_z \dfrac{\partial v_z}{\partial z}\right) = -\dfrac{\partial p}{\partial z}$

$\qquad\qquad\qquad\qquad + \left(\dfrac{1}{r}\dfrac{\partial}{\partial r}(r\tau_{rz}) + \dfrac{1}{r}\dfrac{\partial \tau_{\theta z}}{\partial \theta} + \dfrac{\partial \tau_{zz}}{\partial z}\right) + \rho g_z \quad (C)$

The Continuity Equation:

$$\dfrac{1}{r}\dfrac{\partial}{\partial r}(rv_r) + \dfrac{1}{r}\dfrac{\partial v_\theta}{\partial \theta} + \dfrac{\partial v_z}{\partial z} = 0$$

INTRODUCTION

Table 1.3
Spherical Coordinates (r, θ, ϕ)

The Dynamic Equations:

r-Component
$$\rho\left(\frac{\partial v_r}{\partial t} + v_r\frac{\partial v_r}{\partial r} + \frac{v_\theta}{r}\frac{\partial v_r}{\partial \theta} + \frac{v_\phi}{r\sin\theta}\frac{\partial v_r}{\partial \phi} - \frac{v_\theta^2 + v_\phi^2}{r}\right)$$
$$= -\frac{\partial p}{\partial r} + \left(\frac{1}{r^2}\frac{\partial}{\partial r}(r^2\tau_{rr}) + \frac{1}{r\sin\theta}\frac{\partial}{\partial \theta}(\tau_{r\theta}\sin\theta) + \frac{1}{r\sin\theta}\frac{\partial \tau_{r\phi}}{\partial \phi}\right.$$
$$\left. - \frac{\tau_{\theta\theta} + \tau_{\phi\phi}}{r}\right) + \rho g_r \quad (A)$$

θ-Component
$$\rho\left(\frac{\partial v_\theta}{\partial t} + v_r\frac{\partial v_\theta}{\partial r} + \frac{v_\theta}{r}\frac{\partial v_\theta}{\partial \theta} + \frac{v_\phi}{r\sin\theta}\frac{\partial v_\theta}{\partial \phi} + \frac{v_r v_\theta}{r} - \frac{v_\phi^2 \cot\theta}{r}\right)$$
$$= -\frac{1}{r}\frac{\partial p}{\partial \theta} + \left(\frac{1}{r^2}\frac{\partial}{\partial r}(r^2\tau_{r\theta}) + \frac{1}{r\sin\theta}\frac{\partial}{\partial \theta}(\tau_{\theta\theta}\sin\theta)\right.$$
$$\left. + \frac{1}{r\sin\theta}\frac{\partial \tau_{\theta\phi}}{\partial \phi} + \frac{\tau_{r\theta}}{r} - \frac{\cot\theta}{r}\tau_{\phi\phi}\right) + \rho g_\theta \quad (B)$$

ϕ-Component
$$\rho\left(\frac{\partial v_\phi}{\partial t} + v_r\frac{\partial v_\phi}{\partial r} + \frac{v_\theta}{r}\frac{\partial v_\phi}{\partial \theta} + \frac{v_\phi}{r\sin\theta}\frac{\partial v_\phi}{\partial \phi} + \frac{v_\phi v_r}{r} + \frac{v_\theta v_\phi}{r}\cot\theta\right)$$
$$= -\frac{1}{r\sin\theta}\frac{\partial p}{\partial \phi} + \left(\frac{1}{r^2}\frac{\partial}{\partial r}(r^2\tau_{r\phi}) + \frac{1}{r}\frac{\partial \tau_{\theta\phi}}{\partial \theta} + \frac{1}{r\sin\theta}\frac{\partial \tau_{\phi\phi}}{\partial \phi}\right.$$
$$\left. + \frac{\tau_{r\phi}}{r} + \frac{2\cot\theta}{r}\tau_{\theta\phi}\right) + \rho g_\phi \quad (C)$$

The Continuity Equation:
$$\frac{1}{r^2}\frac{\partial}{\partial r}(r^2 v_r) + \frac{1}{r\sin\theta}\frac{\partial}{\partial \theta}(v_\theta\sin\theta) + \frac{1}{r\sin\theta}\frac{\partial v_\phi}{\partial \phi} = 0$$

One may observe at this stage that the number of unknowns in these equations is much greater than the number of equations at hand, and so more information, in the form of equations, is required. Normally **f** is given in a dynamic problem and ρ is known. Hence, the unknowns are the three components of **v**, the pressure p, and the nine components of $\boldsymbol{\tau}$. Actually, for most real fluids, $\boldsymbol{\tau}$ is a symmetric tensor and has only six independent components.* Hence, there are ten unknowns to be found from only four equations. The additional six equations required are the "constitutive equations" for the material, which relate the components of $\boldsymbol{\tau}$ to the velocity and its derivatives.

For example, a simple constitutive equation is that of the newtonian fluid:

$$\tau_{ij} = \eta_0\left(\frac{\partial v_i}{\partial x_j} + \frac{\partial v_j}{\partial x_i}\right) \quad (1.4)$$

where η_0 is a constant, called the coefficient of viscosity.

* Possible exceptions are so-called polar fluids, or structured continua (1).

Once the constitutive equation is established, the dynamic problem is determinate, but it may not be amenable to solution for complex materials (defined by the constitutive equation) or for complex boundary conditions.

III. RESPONSE OF A FLUID

A fluid responds to stress by flowing. Flow is essentially a process in which the material deforms at a finite rate. The basic kinematic measure of the response of a fluid is the rate of deformation tensor Δ, whose cartesian components are

$$\Delta_{ij} = \frac{\partial v_i}{\partial x_j} + \frac{\partial v_j}{\partial x_i} = \Delta_{ji} \tag{1.5}$$

Table 1.4 gives components of Δ in cartesian, cylindrical, and spherical coordinates.

Motion may exist in a fluid even if Δ is identically zero. Each element of fluid might be translating at the same linear velocity, and each element might, in addition, have the same angular velocity about some axis, due to a rigid rotation of the fluid. But uniform translation and rotation do not contribute to the deformation of the fluid, and so are not associated with that part of the response of the material which is of interest to us.

In addition to a *dynamic* response, real fluids exhibit a *thermodynamic* response. Deformation gives rise to frictional forces within the fluid, and this friction dissipates a part of the kinetic energy of the fluid and causes it to appear as heat. It is possible that sufficient heat is generated to raise the temperature of the fluid appreciably.

One concludes from this that no flow is isothermal, despite any precautions of thermostating the boundaries of the system. However, for relatively low rates of deformation, the temperature rise is insufficient to change the properties of the fluid. On the other hand, for many important flows, the viscosity* of the fluid is so high that even small deformation rates generate significant amounts of heat. In this case one must be able to correct any calculations based on an isothermal analysis. This would require a knowledge of the temperature field in the fluid, as well as a knowledge of the effect of temperature on fluid properties such as viscosity.

The temperature field in an incompressible fluid satisfies an energy equation which, in cartesian coordinates, has the form (2)

$$\rho \hat{C}_p \left(\frac{\partial T}{\partial t} + v_j \frac{\partial T}{\partial x_j} \right) = \frac{\partial}{\partial x_j} \left(k \frac{\partial T}{\partial x_j} \right) + \tau_{ij} \frac{\partial v_i}{\partial x_j} \tag{1.6}$$

* The term viscosity is used here in the general sense of a measure of resistance to deformation.

Table 1.4
Components of the Rate of Deformation Tensor

Cartesian Coordinates (x, y, z):

$$\Delta_{xx} = 2\frac{\partial v_x}{\partial x} \qquad \Delta_{xy} = \Delta_{yx} = \frac{\partial v_x}{\partial y} + \frac{\partial v_y}{\partial x}$$

$$\Delta_{yy} = 2\frac{\partial v_y}{\partial y} \qquad \Delta_{xz} = \Delta_{zx} = \frac{\partial v_x}{\partial z} + \frac{\partial v_z}{\partial x}$$

$$\Delta_{zz} = 2\frac{\partial v_z}{\partial z} \qquad \Delta_{yz} = \Delta_{zy} = \frac{\partial v_y}{\partial z} + \frac{\partial v_z}{\partial y}$$

Cylindrical Coordinates (r, θ, z):

$$\Delta_{rr} = 2\frac{\partial v_r}{\partial r} \qquad \Delta_{r\theta} = \Delta_{\theta r} = r\frac{\partial}{\partial r}\left(\frac{v_\theta}{r}\right) + \frac{1}{r}\frac{\partial v_r}{\partial \theta}$$

$$\Delta_{\theta\theta} = 2\left(\frac{1}{r}\frac{\partial v_\theta}{\partial \theta} + \frac{v_r}{r}\right) \qquad \Delta_{\theta z} = \Delta_{z\theta} = \frac{\partial v_\theta}{\partial z} + \frac{1}{r}\frac{\partial v_z}{\partial \theta}$$

$$\Delta_{zz} = 2\frac{\partial v_z}{\partial z} \qquad \Delta_{zr} = \Delta_{rz} = \frac{\partial v_z}{\partial r} + \frac{\partial v_r}{\partial z}$$

Spherical Coordinates (r, θ, ϕ):

$$\Delta_{rr} = 2\frac{\partial v_r}{\partial r} \qquad \Delta_{\theta r} = \Delta_{r\theta} = r\frac{\partial}{\partial r}\left(\frac{v_\theta}{r}\right) + \frac{1}{r}\frac{\partial v_r}{\partial \theta}$$

$$\Delta_{\theta\theta} = 2\left(\frac{1}{r}\frac{\partial v_\theta}{\partial \theta} + \frac{v_r}{r}\right) \qquad \Delta_{\phi\theta} = \Delta_{\theta\phi} = \frac{\sin\theta}{r}\frac{\partial}{\partial \theta}\left(\frac{v_\phi}{\sin\theta}\right)$$

$$+ \frac{1}{r\sin\theta}\frac{\partial v_\theta}{\partial \phi}$$

$$\Delta_{\phi\phi} = 2\left(\frac{1}{r\sin\theta}\frac{\partial v_\phi}{\partial \phi} + \frac{v_r}{r} + \frac{v_\theta \cot\theta}{r}\right) \qquad \Delta_{r\phi} = \Delta_{\phi r} = \frac{1}{r\sin\theta}\frac{\partial v_r}{\partial \phi} + r\frac{\partial}{\partial r}\left(\frac{v_\phi}{r}\right)$$

The heat capacity per unit mass is $\hat{C}_p$ and the thermal conductivity is k. The term $\tau_{ij}(\partial v_i/\partial x_j) = \Phi$ gives the volumetric rate of conversion of kinetic energy into heat through friction.* It is a dissipation term. Table 1.5 shows Φ in cartesian, cylindrical, and spherical coordinates. Some applications of Eq. 1.6 to the correction of measurements of fluid properties will be given later.

* The function Φ not is the same as the function Φ_v defined by Bird et al. (2).

Table 1.5
The Dissipation Function Φ

Cartesian Coordinates (x, y, z):

$$\Phi = \tau_{xx}\frac{\partial v_x}{\partial x} + \tau_{yy}\frac{\partial v_y}{\partial y} + \tau_{zz}\frac{\partial v_z}{\partial z} + \tau_{xy}\left(\frac{\partial v_x}{\partial y} + \frac{\partial v_y}{\partial x}\right)$$
$$+ \tau_{yz}\left(\frac{\partial v_y}{\partial z} + \frac{\partial v_z}{\partial y}\right) + \tau_{zx}\left(\frac{\partial v_z}{\partial x} + \frac{\partial v_x}{\partial z}\right)$$

Cylindrical Coordinates (r, θ, z):

$$\Phi = \tau_{rr}\left(\frac{\partial v_r}{\partial r}\right) + \tau_{\theta\theta}\left(\frac{1}{r}\frac{\partial v_\theta}{\partial \theta} + \frac{v_r}{r}\right) + \tau_{zz}\left(\frac{\partial v_z}{\partial z}\right)$$
$$+ \tau_{r\theta}\left[r\frac{\partial}{\partial r}\left(\frac{v_\theta}{r}\right) + \frac{1}{r}\frac{\partial v_r}{\partial \theta}\right] + \tau_{\theta z}\left(\frac{1}{r}\frac{\partial v_z}{\partial \theta} + \frac{\partial v_\theta}{\partial z}\right) + \tau_{rz}\left(\frac{\partial v_z}{\partial r} + \frac{\partial v_r}{\partial z}\right)$$

Spherical Coordinates (r, θ, ϕ):

$$\Phi = \tau_{rr}\left(\frac{\partial v_r}{\partial r}\right) + \tau_{\theta\theta}\left(\frac{1}{r}\frac{\partial v_\theta}{\partial \theta} + \frac{v_r}{r}\right) + \tau_{\phi\phi}\left(\frac{1}{r \sin\theta}\frac{\partial v_\phi}{\partial \phi} + \frac{v_r}{r} + \frac{v_\theta \cot\theta}{r}\right)$$
$$+ \tau_{r\theta}\left(\frac{\partial v_\theta}{\partial r} + \frac{1}{r}\frac{\partial v_r}{\partial \theta} - \frac{v_\theta}{r}\right) + \tau_{r\phi}\left(\frac{\partial v_\phi}{\partial r} + \frac{1}{r \sin\theta}\frac{\partial v_r}{\partial \phi} - \frac{v_\phi}{r}\right)$$
$$+ \tau_{\theta\phi}\left(\frac{1}{r}\frac{\partial v_\phi}{\partial \theta} + \frac{1}{r \sin\theta}\frac{\partial v_\theta}{\partial \phi} - \frac{\cot\theta}{r}v_\phi\right)$$

IV. THE SIMPLE SHEAR FLOW

A particularly simple flow which will appear throughout this book is the so-called "simple shear flow," in which there is a non-zero component of velocity in only a single direction. If the subscripts 1, 2, and 3 denote, respectively, the flow direction, the direction of velocity variation, and the neutral direction, then a simple shear flow is defined by*

$$v = (v_1, 0, 0) \tag{1.7}$$

$$\Delta = \dot{\gamma}(x_2)\begin{pmatrix} 0 & 1 & 0 \\ 1 & 0 & 0 \\ 0 & 0 & 0 \end{pmatrix} \tag{1.8}$$

where $\dot{\gamma}(x_2)$ is some scalar function of the x_2-coordinate. $\dot{\gamma}$ is a "shear"

---
* These flows are also called "viscometric flows," because they are achieved in capillary, Couette, and cone and plate viscometers. However, viscometric flows include some non-simple shear flows (3).

component of the rate of deformation tensor, and is usually called the "shear rate."*

The correspondence of the simple shear flow notation to the more common coordinate notation is given in Table 1.6 for some flows of interest. The flows illustrated correspond to the most common flows found in viscometric instruments, but are by no means the only simple shear

Table 1.6

Simple Shear Flow Notation

Flow Geometry	Coordinate Notation		
	1	2	3
1. Poiseuille Flow	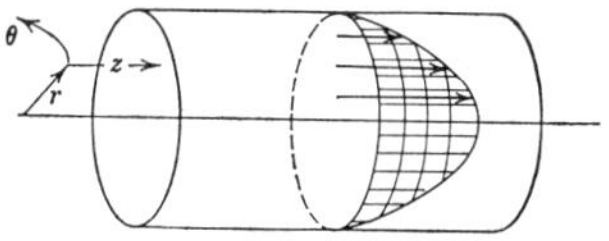 z	r	θ
2. Couette Flow	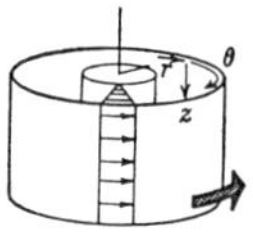 θ	r	z
3. Parallel Plate Torsion	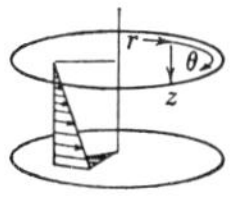 θ	z	r
4. Cone and Plate Torsion	ϕ	θ	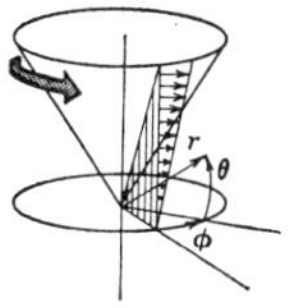 r

flows. According to these definitions, the only non-zero components of $\mathbf{\Delta}$ in a simple shear flow are Δ_{12} and Δ_{21} ($=\Delta_{12}$), as required by Eq. 1.8.

The stress components of a simple shear flow are labeled in the same

* In subsequent discussions $\dot{\gamma}$ is always taken as a *positive* quantity and so represents the *magnitude* of Δ_{12}.

manner. Thus τ_{12} is the shearing stress and τ_{11} is the normal stress in the direction of flow.

One now defines the "apparent viscosity" η of a fluid as the ratio of the shearing stress to the shear rate, or

$$\eta = \tau_{12}/\Delta_{12} = \tau_{12}/\dot{\gamma} \tag{1.9}$$

This definition is consistent with the usual definition of the viscosity of a newtonian fluid, and serves as a definition of the viscosity of a non-newtonian fluid. Chapter 2 considers the analysis of simple shear flows which allow evaluation of the terms in Eq. 1.9, thus leading to the measurement of viscosity.

In addition to shear stresses, a simple shear flow may be accompanied by normal stresses. It is common to define a normal stress coefficient as

$$\Psi_{ii} = \tau_{ii}/\dot{\gamma}^2 \tag{1.10}*$$

In addition, one often considers stress differences, in which case a coefficient

$$\Psi_{ij} = (\tau_{ii} - \tau_{jj})/\dot{\gamma}^2 \tag{1.11}$$

may be defined.

This particular definition of a normal stress coefficient is taken because many fluids exhibit a quadratic dependence of normal stresses on shear rate in the limit of low shear rate. For such fluids Ψ would approach a constant value Ψ_0 characteristic of the fluid. This parallels the observation that many non-newtonian fluids exhibit a linear dependence of τ_{12} on $\dot{\gamma}$ in the limit of small $\dot{\gamma}$, and so η approaches a constant value η_0 (the "zero-shear" viscosity) characteristic of the fluid. Chapter 2 describes a number of techniques whereby Ψ may be measured.

V. SIMPLE ELONGATION

In addition to shear flows one may produce elongational flows, defined by a rate of deformation tensor which has only diagonal components:

$$\Delta = \begin{pmatrix} \Delta_{11} & 0 & 0 \\ 0 & \Delta_{22} & 0 \\ 0 & 0 & \Delta_{33} \end{pmatrix} \tag{1.12}$$

Simple elongation may be defined by a rate of deformation tensor

$$\Delta = \dot{\epsilon} \begin{pmatrix} 2 & 0 & 0 \\ 0 & -1 & 0 \\ 0 & 0 & -1 \end{pmatrix} \tag{1.13}$$

where $\dot{\epsilon}$ is a constant, called the strain rate.

* No summation is implied by the repeated subscripts in Eqs. 1.10 and 1.11.

INTRODUCTION

Elongational flows arise in such important polymer processes as fiber spinning and film drawing. The stress accompanying the extension of the material is of particular importance in defining operating conditions for such processes. For this reason one defines an elongational viscosity η_e (sometimes called a tensile viscosity) as

$$\eta_e = T_{11}/\dot{\epsilon} \tag{1.14}$$

Note that η_e is defined in terms of the total stress T_{11} and not in terms of the dynamic stress τ_{11}.

The elongational viscosity, like the shear viscosity, may be a function of the rate of deformation. For the newtonian fluid, however, one may establish a very simple relationship between η_0 and η_e. Consider an elongational flow established by pulling one end of a cylindrical rod of fluid. (In order to maintain its shape such a fluid would have to be extremely viscous and have a viscosity of the order of 10^6 poise.) Let the rate of deformation tensor be given by Eq. 1.13, that is, assume that a simple elongation has been established. (Experimental aspects of this problem will be discussed in Chapter 2, p. 60.) For the newtonian fluid one has $\tau_{ij} = \eta_0 \Delta_{ij}$, and the total stress is $T_{ij} = -p\delta_{ij} + \tau_{ij}$. This gives

$$\tau_{11} = 2\eta_0 \dot{\epsilon}$$
$$\tau_{22} = \tau_{33} = -\dot{\epsilon}\eta_0 = -\tfrac{1}{2}\tau_{11} \tag{1.15}$$

If the definition of pressure is taken as $p = -\tfrac{1}{3}T_{ii}$, then

$$p = -\tfrac{1}{3}(T_{11} + T_{22} + T_{33}) = -\tfrac{1}{3}(-p + \tau_{11} + T_{22} + T_{33}) \tag{1.16}$$

If no forces act on the cylindrical face of the rod* then the stresses T_{22} and T_{33} are zero. It follows, from Eq. 1.16, that $p = -\tfrac{1}{2}\tau_{11}$. Hence the total axial stress is

$$T_{11} = -p + \tau_{11} = \tfrac{3}{2}\tau_{11} = 3\eta_0\dot{\epsilon} \tag{1.17}$$

and the elongational viscosity is seen to be

$$\eta_e = 3\eta_0 \quad \text{for a newtonian fluid} \tag{1.18}$$

This result is usually attributed to Trouton (4), and η_e is sometimes referred to as the Trouton viscosity.

In addition to the flows introduced in this chapter, a number of more complex, usually transient, deformation programs are used in the complete investigation of material response. The simple flows described already, as well as these more complicated deformations, will be described in some detail in Chapter 2.

---

* Interfacial tension σ would be a source of radial stress, and T_{22} would be given by $T_{22} = \sigma/R$, where R is the radius of the cylinder. This stress is usually negligible in comparison to dynamic stresses.

REFERENCES

1. Dahler, J. S., and L. E. Scriven, *Nature*, **192**, 36 (1961).
2. Bird, R. B., W. E. Stewart, and E. N. Lightfoot, *Transport Phenomena*, Wiley, New York, 1960.
3. Coleman, B. D., and W. Noll, *Arch. Ratl. Mech. Anal.*, **2**, 197 (1958).
4. Trouton, F. T., *Proc. Roy. Soc.*, **A77,** 426 (1906).

2

Experimental Techniques

In this chapter the problems involved in measuring fundamental rheological parameters are to be discussed. No attempt is made to describe in detail the design and operation of equipment suitable for such measurement. The reader interested in such matters should consult the book by Van Wazer *et al.* (1). Instead, the general analyses of those idealized flows upon which most measuring devices are based are to be presented. Limitations which prevent the complete realization of a simple shear flow in practice, and methods of treating rheological data from which the described properties are to be extracted, will be discussed.

To test the validity of a postulated constitutive equation, it is necessary to perform an experiment in which the variation of components of the stress tensor with components of the rate of deformation tensor can be determined. In principle one must have at hand a solution to the dynamic equations which describe the flow of interest. If the solution is approximate it must be subject to error smaller than the acceptable error of measurement. This is the primary reason that accurate and reliable experiments are based upon idealized flows which yield to correspondingly simple mathematical formulation and solution, such as the simple shear flows, or simple elongation. But therein lies a limitation of the generality of information generated by any such simple shear flow, for such a flow gives information about the dependence of the measured stresses upon only *one* component of the rate of deformation tensor, while a constitutive equation purports to be a relationship involving *all* components of the rate of deformation tensor. Hence, simple flows can only partially establish a constitutive equation, and more than one type of flow is required to give more confidence in the generality of the information developed.

I. STEADY STATE FLOWS
A. Viscosity
1. Poiseuille Flow—The Capillary Viscometer

The constitutive equations are *microscopic* equations in the sense that they define the stress–rate of deformation relationship at every point in the fluid. On the other hand one is usually only capable of easily measuring *macroscopic* variables; for example, volume flow rate can be measured more easily than the velocity components at every point in the fluid. Hence a method of writing the functional dependence implied by a general constitutive equation is sought in terms of macroscopically measurable variables.

Consider a laminar steady state flow through a circular tube or capillary of radius R. The volume flow rate Q is related to the velocity $v_z(r)$ by

$$Q = \int_0^R 2\pi r v_z(r) \, dr = \pi R^2 \langle V \rangle \tag{2.1}$$

where $\langle V \rangle$ denotes the mass average velocity. An integration by parts yields

$$Q = \pi r^2 v_z \big|_0^R - \int_0^R \pi r^2 \frac{dv_z}{dr} \, dr \tag{2.2}$$

If a "no-slip" condition holds at R, i.e., $v_z = 0$ at $r = R$, then

$$Q = -\int_0^R \pi r^2 \frac{dv_z}{dr} \, dr \tag{2.3}$$

For this simple shear flow the only non-zero component of the rate of deformation tensor is $dv_z/dr = \Delta_{12}$. Now let the shear rate–shear stress relationship be written in the functional notation

$$-\Delta_{12} = f(\tau_{12}) \quad \text{or simply} \quad \dot{\gamma} = f(\tau) \tag{2.4}$$

where τ is the shear stress component. From the dynamic equations for this flow it can be shown that the shear stress is linear across the tube radius, or (2, p. 45)

$$\tau = C_0 r/2 \tag{2.5}$$

where $C_0 = \Delta P/L$ is the pressure gradient causing flow.

The wall shear stress is given by

$$\tau_w = R \Delta P/2L \tag{2.6}$$

From Eq. 2.5 it follows that

$$d\tau = C_0 \, dr/2 \tag{2.7}$$

and

$$r^2 = 4\tau^2/C_0^2 \tag{2.8}$$

If these results are substituted into Eq. 2.3, one finds

$$\frac{4Q}{\pi R^3} = \phi(\tau_w) = \frac{4}{\tau_w^3} \int_0^{\tau_w} \tau^2 f(\tau) \, d\tau \tag{2.9}$$

where $\phi(\tau_w)$ is defined by this equation.

EXPERIMENTAL TECHNIQUES

Taking $d\phi/d\tau_w$, and using Leibniz' rule (3), one finds

$$\frac{d\phi}{d\tau_w} = -\frac{12}{\tau_w^4}\int_0^{\tau_w} \tau^2 f(\tau)\,d\tau + \frac{4}{\tau_w}f(\tau_w) \qquad (2.10)$$

or, using Eq. 2.9

$$\frac{d\phi}{d\tau_w} = -\frac{3\phi}{\tau_w} + \frac{4}{\tau_w}f(\tau_w) \qquad (2.11)$$

Finally, solving for $f(\tau_w)$, one finds

$$f(\tau_w) = \tfrac{3}{4}\phi + \tfrac{1}{4}\tau_w \frac{d\phi}{d\tau_w} \qquad (2.12)$$

Equation 2.12 is known as the Weissenberg-Rabinowitsch-Mooney equation.

Since ϕ and τ_w are measurable in terms of macroscopic variables, Eq. 2.12 allows one to find $f(\tau_w)$ over some range of τ_w. But this functional dependence is identical with $f(\tau)$, so, with Eq. 2.4, the shear behavior of the fluid is determined. The usual method is to plot ϕ vs. τ_w, and from this obtain $d\phi/d\tau_w$. Then the right-hand side of Eq. 2.12 is plotted against τ_w to give $f(\tau_w)$. From the resulting graph of $f(\tau)$ vs. τ, which is just Δ_{12} vs. τ_{12}, one can obtain and plot the viscosity η as a function of shear rate through the definition in Eq. 1.9.

This simple analysis of flow in a capillary is based upon the assumption that a simple shear flow exists. This is achieved in steady state, laminar, isothermal flow in a tube of constant cross section, as long as one does not consider regions near the entrance and exit of the tube. In these "end" regions the flow is changing from (or to) its previous (or future) distribution outside the tube. The length of an "end" region is generally a function of tube diameter and some dynamic parameters. For example, for a newtonian fluid, the "entrance length," the length of tube required to achieve the fully developed simple shear flow, depends upon the Reynolds number (2, p. 47)

$$L_e/D = 0.035\frac{\langle V\rangle D\rho}{\eta} = 0.035\,\mathrm{Re} \qquad (2.13)$$

where D is the tube diameter.

One method of minimizing the effect of the entrance length is to use a viscometer tube so long that the pressure drop over the entrance region is very small compared with the drop over the entire tube. Generally this means that if L is the total tube length, then L_e/L must be small, perhaps of the order of 0.01. Often this restriction may demand tubes too long for practical purposes.

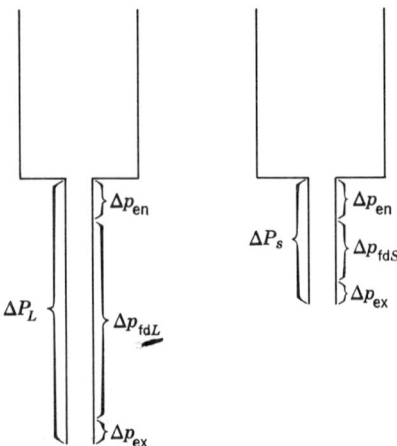

Fig. 2.1. Definition sketch for the treatment of capillary flow data.

A method for eliminating the influence of end effects exists which uses two tubes identical except in length. The shorter tube must be longer than the entrance length, so that both tubes have a finite region in which the flow is in simple shear.

Consider two such tubes attached to the bottoms of identical pressurized reservoirs, as indicated in Fig. 2.1. Imagine that the pressure drops across each tube are adjusted so that the volume flow rates are identical. The reservoirs are assumed to contain identical fluids.

Let ΔP_L and ΔP_S be the total pressure drops measured across the long and short tubes, respectively. If the volume flow rate is the same in both tubes, then one would expect the entrance and exit lengths and the pressure drops across these lengths to be identical in both tubes. Then we may write

$$\Delta P_L = \Delta p_{en} + \Delta p_{ex} + \Delta p_{fdL} \tag{2.14}$$

$$\Delta P_S = \Delta p_{en} + \Delta p_{ex} + \Delta p_{fdS} \tag{2.15}$$

where Δp_{en} and Δp_{ex} are the pressure drops across the entrance and exit regions and Δp_{fd} is the pressure drop across the long (L) or short (S) tube that arises strictly from the fully developed flow. Since the flow rates are the same, the pressure *gradients* in the fully developed sections are identical, and

$$\Delta p_{fdL}/L_L = \Delta p_{fdS}/L_S \tag{2.16}$$

where L_L and L_S are the lengths over which the flow is fully developed in

EXPERIMENTAL TECHNIQUES

Table 2.1
Pressure–Flow Rate Data for 1.5% CMC

D (cm.)	L (cm.)	ΔP (psi)	Q (cm.^{3}/sec.)	τ_w (dynes/cm.^{2})	ϕ (sec.$^{-1}$)
0.271	94.4	20.0	0.591	994	302
		39.3	2.95	1,950	1,510
		59.5	8.32	2,960	4,260
		85.8	21.0	4,250	10,700
0.271	67.4	15.0	0.733	1,040	375
		20.0	1.26	1,390	645
		39.5	7.06	2,750	3,620
		58.0	17.6	4,040	9,030
		79.5	35.1	5,540	17,950
0.182	63.4	20.5	0.192	1,020	324
		28.0	0.477	1,390	805
		37.5	0.801	1,860	1,350
		59.0	2.58	2,920	4,360
		79.5	5.56	3,940	9,400
		100.3	9.34	5,000	15,800
0.182	45.2	9.0	0.067	625	112
		20.5	0.350	1,420	591
		31.0	0.972	2,160	1,645
		45.0	2.72	3,120	4,600
		59.5	5.75	4,140	9,720
		75.0	9.84	5,200	16,600
0.0856	29.7	41.0	0.0906	2,040	1,470
		81.0	0.581	4,030	9,450
		100.0	1.00	4,970	16,250
		126.5	1.74	6,300	28,300
		154.0	2.66	7,650	43,300
0.0856	20.95	31.3	0.108	2,210	1,760
		55.0	0.506	3,880	8,250
		70.0	0.947	4,950	15,400
		91.5	1.81	6,450	29,400
		113.5	3.20	8,000	52,000
		134.0	3.85	9,460	62,600

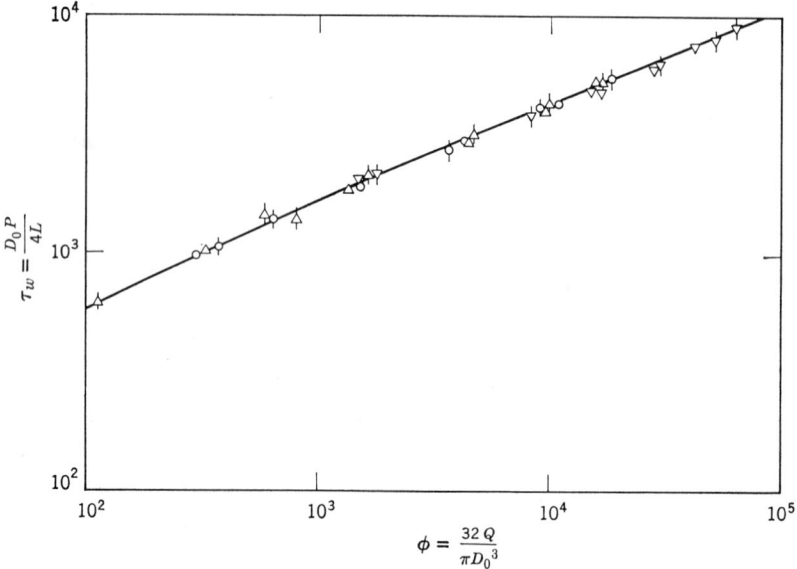

Fig. 2.2. Capillary flow data for an aqueous solution of carboxymethylcellulose (CMC). The data are not corrected for entrance effects.

the large and small tubes. If Eq. 2.15 is subtracted from Eq. 2.14 and Eq. 2.16 is used, the result is (4)

$$\frac{\Delta p_{\text{fd}L}}{L_L} = \frac{\Delta P_L - \Delta P_S}{L_L - L_S} \tag{2.17}$$

Equation 2.17 gives a pressure gradient free of entrance and exit effects, although either viscometer tube may have involved appreciable end effects. It should be noted that the tube lengths should be sufficiently different so that the pressure difference $\Delta P_L - \Delta P_S$ is not subject to large error.

As an illustration of the techniques discussed here consider the data presented in Table 2.1. Since the minimum L/D for these data is 244, no attempt was made to correct for entrance and exit effects. Figure 2.2 shows the results plotted as τ_w vs. ϕ. Two points should be noted. The first is that the data fall on a straight line, which can be written as

$$\tau_w = K'\phi^n \tag{2.18}$$

From Eq. 2.12 one finds

$$\dot{\gamma} = f(\tau) = \left(\frac{3n+1}{4n}\right)\left(\frac{\tau}{K'}\right)^{1/n} \tag{2.19}$$

which can be written as

$$\tau = \left(\frac{4n}{3n+1}\dot\gamma\right)^n K' \qquad (2.20)$$

This expression can be put into the standard form for the power law, $\tau = K\dot\gamma^n$, where

$$K = K'\left(\frac{4n}{3n+1}\right)^n \qquad (2.21)$$

For this example, $n = 0.4$ and $K = 97$ dynes cm.$^{-2}$-sec.$^{0.4}$.

The second point is the observation that the data for the shorter tubes appear to be no different from the data for the longer tubes. This indicates that the error in neglecting entrance and exit effects is masked by any experimental error, justifying our failure to correct ΔP, and so τ_w. This is to be expected for tubes as long as those used here.

2. Couette Flow—The Coaxial Cylinder Viscometer

A common type of rheological instrument consists of two coaxial cylinders, with the fluid to be studied held in the annular space between. Mechanical construction allows angular motion of one of the cylinders. A means is provided for measuring the torque exerted on or by the cylinders. Figure 2.3 shows a schematic drawing of such an instrument.

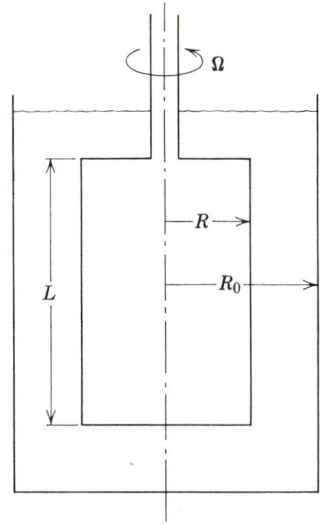

Fig. 2.3. Definition sketch for the treatment of coaxial cylinder flow data.

One assumes that a simple shear flow exists in the annular region, such that
$$\mathbf{v} = (v_\theta, 0, 0) \tag{2.22}$$
and
$$v_\theta = v_\theta(r) = r\omega(r) \tag{2.23}$$
where ω is the angular velocity (radians/sec.).

The rate of deformation tensor for this simple shear flow has only components
$$\Delta_{12} = r\frac{d\omega}{dr} = -\dot{\gamma} \tag{2.24}$$

Define a function
$$f(\tau) = -r\frac{d\omega}{dr} = \dot{\gamma} \tag{2.25}$$

where τ is the shear stress component τ_{12}.

Assume that the outer cylinder is stationary and that the inner cylinder is driven with an angular velocity Ω by the application of a torque $\mathcal{T}$. This torque must balance the torque exerted by the fluid on the face of the inner cylinder, so that
$$\mathcal{T} = 2\pi R^2 L \tau_R \tag{2.26}$$

where L and R are the height and radius of the inner cylinder, and τ_R is the shear stress exerted on the inner cylinder.

For a simple shear flow, the dynamic equations lead immediately to
$$r^2 \tau = \text{constant} \tag{2.27}$$

from which one finds
$$r^2\,d\tau + 2r\tau\,dr = 0 \tag{2.28}$$

This allows the transformation
$$\frac{d}{dr} = -\frac{2\tau}{r}\frac{d}{d\tau} \tag{2.29}$$

and so Eq. 2.25 becomes
$$f(\tau) = 2\tau\frac{d\omega}{d\tau} \tag{2.30}$$

It follows that
$$\omega = \frac{1}{2}\int_{\tau_0}^{\tau}\frac{f(\tau)}{\tau}\,d\tau \tag{2.31}$$

when the boundary condition

$$\omega = 0 \quad \text{at} \quad r = R_0 \tag{2.32}$$

is used and the shear stress at the outer cylinder is designated by τ_0.
Equation 2.31 is valid, in particular at $r = R$, and becomes

$$\Omega = \frac{1}{2} \int_{\tau_0}^{\tau_R} \frac{f(\tau)}{\tau} d\tau \tag{2.33}$$

Now differentiate Eq. 2.33 with respect to τ_R and find

$$\frac{d\Omega}{d\tau_R} = \frac{1}{2} \left[\frac{f(\tau_R)}{\tau_R} - \frac{f(\tau_0)}{\tau_0} \frac{d\tau_0}{d\tau_R} \right] \tag{2.34}$$

But, from Eq. 2.27

$$\tau_0 R_0^2 = \tau_R R^2 \tag{2.35}$$

and hence

$$\frac{d\tau_0}{d\tau_R} = \left(\frac{R}{R_0}\right)^2 = s^2 \tag{2.36}$$

Thus Eq. 2.34 becomes

$$2\tau_R \frac{d\Omega}{d\tau_R} = f(\tau_R) - f(s^2 \tau_R) \tag{2.37}$$

Since Eq. 2.26 gives τ_R as a function of $\mathscr{T}$, the terms which appear in Eq. 2.37 are determined from Ω vs. $\mathscr{T}$ data. But in its present form, Eq. 2.37 is a *difference* equation for $f(\tau_R)$. In the special case of $s^2 = 0$, corresponding to measurement in an infinite body of fluid, Eq. 2.37 can be reduced to

$$2\tau_R \frac{d\Omega}{d\tau_R} = f(\tau_R) \tag{2.38}$$

since $f(0) = 0$. If $\mathscr{T}$ is introduced into this result, one finds

$$f(\tau_R) = 2\Omega \frac{d \ln \Omega}{d \ln \mathscr{T}} \tag{2.39}$$

Equation 2.39 would be a good approximation for $s < 0.1$.

The difference equation can be solved by an iterative procedure. Equation 2.37 is valid for all values of τ_R. For example, one may write $s^2 \tau_R$ for τ_R and find

$$2s^2 \tau_R \frac{d\Omega}{d\tau_R}\bigg|_{s^2 \tau_R} = f(s^2 \tau_R) - f(s^4 \tau_R) \tag{2.40}$$

and, in general

$$2s^{2(N-1)} \tau_R \frac{d\Omega}{d\tau_R}\bigg|_{s^{2(N-1)} \tau_R} = f(s^{2(N-1)} \tau_R) - f(s^{2N} \tau_R) \tag{2.41}$$

If the complete set of equations obtained by successively substituting $s^2\tau_R, s^4\tau_R, \ldots, s^{2(N-1)}\tau_R$ for τ_R in Eq. 2.37 is added together, the result is

$$f(\tau_R) - f(s^{2N}\tau_R) = \sum_{p=0}^{N} 2s^{2p}\tau_R \left.\frac{d\Omega}{d\tau_R}\right|_{s^{2p}\tau_R} \quad (2.42)$$

Now let N go to infinity, and take note of the fact that s^{2N} vanishes in this limit since $s < 1$. The result is (5)

$$f(\tau_R) = \dot\gamma = 2\tau_R \sum_{p=0}^{\infty} s^{2p}\Omega'\Big|_{s^{2p}\tau_R} \quad (2.43)$$

where $\Omega' = d\Omega/d\tau_R$.

It is common practice to plot $\log \Omega$ vs. $\log \tau_R$, from which the slope

$$m = \frac{d \log \Omega}{d \log \tau_R} \quad (2.44)$$

can be more accurately found than the slope Ω', since m is nearly a constant over a wide range of τ_R for many fluids. In that case one calculates the shear rate at the inner cylinder using

$$f(\tau_R) = 2\Omega \sum_{p=0}^{\infty} s^{2pm} m(s^{2p}\tau_R) \quad (2.45)$$

Since s is less than unity, this series converges, but does so slowly in the case that s is nearly unity. This case is of some importance since many instruments of the coaxial type employ an annular space which is thin compared with the cylinder radii, leading to values of s within a few percent of unity. In this case an approximation to Eq. 2.37 can be found, and the result is (6)

$$f(\tau_R) = \frac{\Omega}{-\ln s}[1 - m \ln s + \tfrac{1}{3}(m \ln s)^2] \quad (2.46)$$

This approximation is quite accurate as long as $-m \ln s < 0.5$.

To illustrate the calculations outlined here with a simple example, Table 2.2 shows data obtained with an Epprecht RM-15 Viscometer to measure the properties of a 1% solution of CMC. The Epprecht is a concentric cylinder viscometer in which a cylindrical bob (with conical ends) rotates in a cylindrical cup. The instrument reading can be converted to shear stress using a conversion factor given for each bob. Readings are taken over a series of rotational speeds, and thus over a range of shear rates. A manual supplied with the instrument gives a table of nominal shear rates at each rotational speed, for each cup and bob combination.

For the "B-System," the gap between cylinders is 4 mm., and s is 0.794.

EXPERIMENTAL TECHNIQUES

Table 2.2
Epprecht RM-15 Viscometer "B-System"
1% CMC (70-S Hercules) at 25.0°C.

τ_R (dynes/cm.^{2})	Ω (rpm)	m
41	25.1	1.27
52	33.7	1.27
65	44.3	1.27
81	59.1	1.27
101	78.0	1.27
136	113	1.27
170	152	1.27
207	200	1.46
254	267	1.46
309	352	1.46

Since s is not close to zero or unity, the series correction given by Eq. 2.45 must be used.

As the first step, a double logarithmic plot, Fig. 2.4, of Ω vs. τ_R is made. From this the slope $d \log \Omega / d \log \tau_R = m$ may be obtained for any value

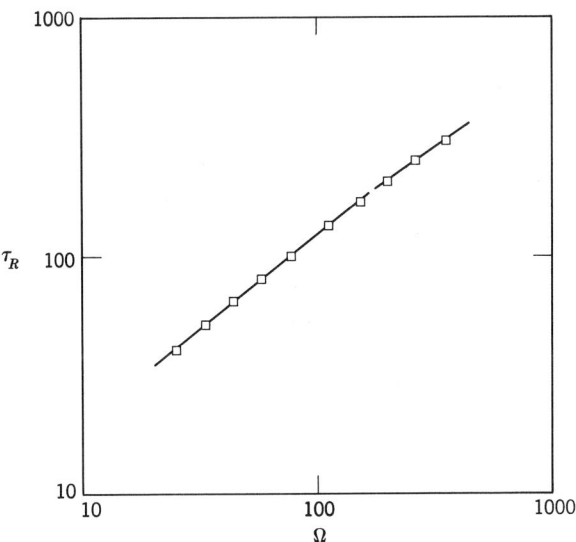

Fig. 2.4. Data of Table 2.2.

of τ_R. For simplicity this curve was fitted with two straight lines, and the values of the slopes are shown tabulated.

To evaluate the shear rate, for example, at $\Omega = 352$ rpm, for which τ_R is 309 dynes/cm.², one must find

$$f(\tau) = \dot{\gamma} = \left(\frac{2\pi}{60}\right) 2(352) \sum_{p=0}^{\infty} (0.794)^{2pm} m[(0.794)^{2p}(309)]$$

where the factor $2\pi/60$ converts rpm to radians/sec. For convenience only seven terms of the series are used. The lowest stress for which a slope is required is $s^{12}\tau_R = 18.6$ dynes/cm.². No data exist for this stress, and one must extrapolate slightly and assume that the slope is unchanged. Thus the last few terms of the series will be in error (since the slope is not exactly known) but the relative contribution of these terms is small, and the total error should be small. The resulting value of $\dot{\gamma}$ is found to be $\dot{\gamma} = 268$ sec.$^{-1}$. This process is continued for the other points in the Table.

It should be clear that one must extrapolate about an order of magnitude below the lower data points in order to find $\dot{\gamma}$ accurately at the low end of the curve. These calculations have been made assuming that the slope remained at 1.27 in the low stress region. If possible, one should take data at stresses below the region of interest in order to make accurate calculations of the shear rate.

Figure 2.5 shows a shear stress–shear rate curve, of which the data of this illustration are a part. The high shear rate data were taken with a capillary viscometer, and $\dot{\gamma}$ was calculated from the Rabinowitsch-Mooney equation. The Epprecht data were plotted with the shear rates calculated from Eq. 2.48. The Epprecht data are in good agreement with the capillary data, as they should be if the proper calculations are carried out, and if one has indeed measured a material property.

For comparison, a line labeled "uncorrected" is shown, for which the shear rates were taken from the Epprecht Service Manual. These shear rates were calculated from

$$\dot{\gamma} = 2\Omega/(1 - s^2) \quad (2.47)$$

which is exact only for the newtonian fluid. The difference between the two curves is not so great in this case, but can be quite significant in practical situations.

For the power law fluid, m of Eq. 2.44 is a constant ($m = 1/n$), and Eq. 2.45 reduces to

$$\dot{\gamma} = \frac{2m\Omega}{1 - s^{2m}} \quad (2.48)$$

By solving the dynamic equations for this flow, for a power law fluid, the same result is found. Equation 2.48 may be used to calculate the shear rate

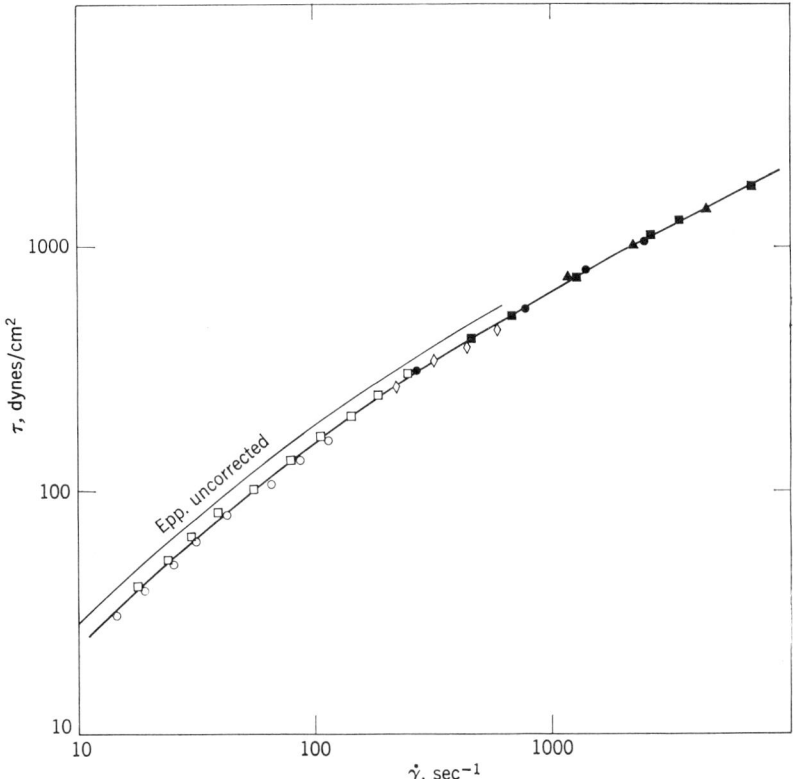

Fig. 2.5. Comparison of data obtained with the same fluid in a capillary viscometer and a coaxial cylinder (Epprecht) viscometer. Both sets of data corrected, as described in the text, for non-newtonian effects. Neither set of data is corrected for end effects. The line labeled "uncorrected" is based on the newtonian shear rate, Eq. 2.47.

whenever the power law is a good approximation. By comparison with Eq. 2.47 it is clear that

$$g(m, s) = m \frac{1 - s^2}{1 - s^{2m}} \qquad (2.49)$$

represents the error made in not correcting coaxial cylinder data for non-newtonian effects. Figure 2.6 shows this function.

3. The Cone and Plate Viscometer (7)

Consider the flow between a rotating cone and a flat plate, as shown in Fig. 2.7. The angle between the solid surfaces ψ_0 is very small in commercial instruments, being generally in the neighborhood of one to five degrees.

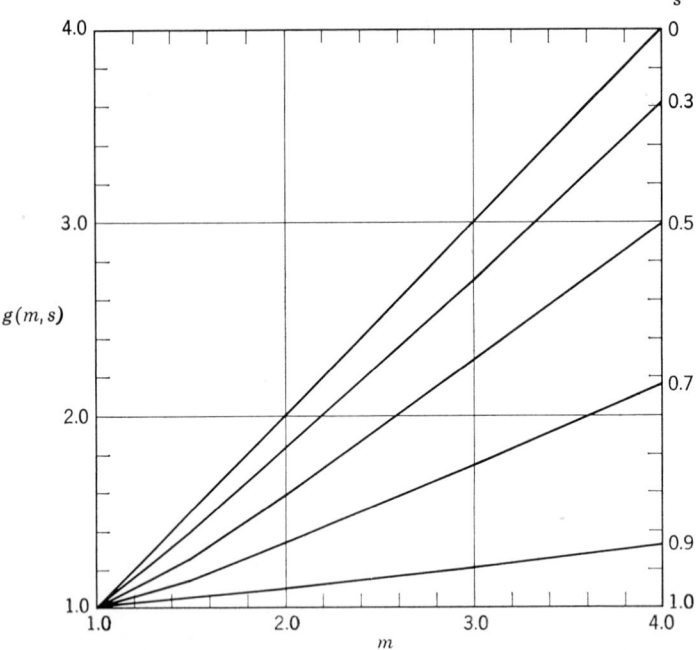

Fig. 2.6. Shear rate error (Eq. 2.49) if coaxial cylinder data are not corrected for non-newtonian effect.

If inertial forces are neglected and edge effects at the periphery of the cone are ignored, the dynamic equation, in spherical coordinates, is

$$\frac{1}{r}\frac{\partial}{\partial r}(r^2 \tau_{r\phi}) + \frac{\partial \tau_{\theta\phi}}{\partial \theta} + \tau_{r\phi} + 2\tau_{\theta\phi} \cot\theta = 0 \qquad (2.50)$$

With the above assumptions, the only non-zero velocity component is v_ϕ, and, by the symmetry of the flow in the ϕ-direction, the only non-zero

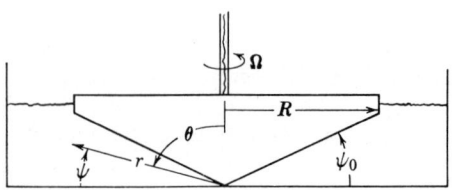

Fig. 2.7. Definition sketch for cone and plate viscometer.

components of the rate of deformation tensor are

$$\Delta_{13} = \Delta_{\phi r} = r \frac{\partial}{\partial r}\left(\frac{v_\phi}{r}\right) \tag{2.51}$$

$$\Delta_{12} = \Delta_{\phi\theta} = \frac{\sin\theta}{r}\frac{\partial}{\partial\theta}\left(\frac{v_\phi}{\sin\theta}\right) \tag{2.52}$$

The boundary conditions for this flow are

$$v_\phi = \Omega r \sin\left(\frac{\pi}{2} - \psi_0\right) \quad \text{at} \quad \psi = \psi_0 \tag{2.53}$$

$$v_\phi = 0 \quad \text{at} \quad \psi = 0 \tag{2.54}$$

$$v_\phi = 0 \quad \text{at} \quad r = 0 \tag{2.55}$$

These conditions suggest a solution of the form

$$v_\phi = rg(\psi) \tag{2.56}$$

This, substituted into the expression for Δ_{13} given above, leads immediately to the conclusion that Δ_{13} vanishes. Furthermore, the symmetry of the flow indicates that the shear stress $\tau_{13} = \tau_{\phi r}$ vanishes.

The dynamic equation now becomes

$$\frac{1}{r}\frac{d\tau_{12}}{d\theta} - \frac{2\cot\theta}{r}\tau_{12} = 0 \tag{2.57}$$

and the solution

$$\tau = \tau_{12} = \frac{C_1}{\sin^2\theta} \tag{2.58}$$

follows immediately. The constant C_1 is obtained from the condition that the measured torque $\mathcal{T}$ arises from the shear stress exerted on the surface of the cone. This leads to

$$\mathcal{T} = \int_0^R \tau_{12}\bigg|_{\pi/2-\psi_0} 2\pi\left[r\sin\left(\frac{\pi}{2}-\psi_0\right)\right]^2 dr \tag{2.59}$$

and, introducing Eq. 2.58, one finds

$$C_1 = 3\mathcal{T}/2\pi R^3 \tag{2.60}$$

From this point one proceeds on the basis of some approximations which depend upon the fact that ψ is very small. From Eq. 2.58 one sees that τ is nearly constant, and, since $\sin\theta = \sin(\pi/2 - \psi) = 1$ approximately

$$\tau = C_1 \tag{2.61}$$

For ψ less than one deg., the error is less than 0.1%. Since τ is nearly constant across the gap, it follows that Δ_{12} is also nearly constant. Within these approximations, Eq. 2.52 becomes

$$-\dot\gamma = \Delta_{12} = \frac{1}{r}\frac{dv_\phi}{d\theta} \tag{2.62}$$

It follows immediately that

$$v_\phi = -r\dot\gamma\left(\theta - \frac{\pi}{2}\right) = r\dot\gamma\psi \tag{2.63}$$

From the boundary condition at $\psi = \psi_0$, the shear rate $\dot\gamma$ may be calculated as

$$\dot\gamma = \Omega/\psi_0 \tag{2.64}$$

and the velocity profile is seen to be

$$v_\phi = r\Omega\psi/\psi_0 \tag{2.65}$$

One sees, then, that for small enough gaps, a plot of $\tau = 3\mathcal{T}/2\pi R^3$ vs. $\dot\gamma = \Omega/\psi_0$ establishes the relationship between τ and $\dot\gamma$. Herein lies the primary advantage of the cone and plate viscometer, for, with good approximation, the shear stress and shear rate are constant throughout the sample, and no graphical differentiation of data is required to obtain the τ vs. $\dot\gamma$ curve.

4. Viscous Heat Generation

The generation of heat through the action of viscous dissipation can lead to significant temperature variations across the shear fields in any of the viscometer configurations discussed in the preceding sections. Because fluid properties such as viscosity are rather strongly temperature dependent the shear stress–shear rate relation is considerably altered by non-isothermal effects. Hence a shear stress–shear rate curve obtained under non-isothermal conditions does not reflect the basic fluid response independently of any temperature-dependent effects. The task at hand is to take data subject to viscous heating effects and separate that part of the response due to non-isothermal behavior from that part of the response due to non-newtonian behavior.

The analyses of the preceding sections are applicable (with minor alterations) to non-isothermal flows. However, these analyses do not allow a separation of the temperature-dependent behavior from the overall response of the fluid. This can be done only if a model for the fluid (a constitutive equation) is assumed, and if the temperature variation of the fluid parameters is specified. This results in a parameter fitting procedure in which data taken at different shear rates and temperatures are manipulated until a consistent set of fluid constants (independent of shear rate and temperature) is obtained.

It would be logical to parallel the earlier discussions with analyses of viscous dissipation in Poiseuille, Couette, and cone and plate simple shear flows. This shall be done, but the results for Poiseuille flow will be seen to be disappointing; the problem just does not lend itself to the simpler analyses possible in the Couette and cone and plate systems.

a. Poiseuille Flow. If heating is important then one must solve the energy equation for laminar flow in cylindrical coordinates. With appropriate boundary conditions the problem to be solved is

$$\rho \hat{C}_p v_z \frac{\partial T}{\partial z} = \frac{1}{r}\frac{\partial}{\partial r}\left(rk\frac{\partial T}{\partial r}\right) + \Phi \qquad (2.66)$$

$$T = T_0 \quad \text{at} \quad z \leqslant 0$$

$$\frac{\partial T}{\partial r} = 0 \quad \text{at} \quad r = 0$$

$$\left.\begin{array}{c} -k\dfrac{\partial T}{\partial r} = q \\ \text{or} \\ T = T_0 \end{array}\right\} \quad \text{at} \quad r = R$$

Here T_0 would be the reservoir temperature of the fluid, and k, ρ, and $\hat{C}_p$ are the fluid thermal conductivity, density, and heat capacity, respectively. The third boundary condition is arbitrary, in that it depends upon actual operating conditions. Typical conditions used state that the heat flux across the tube wall may be a constant q (including zero, the adiabatic condition), or the temperature of the wall may be specified as a constant.

The energy equation 2.66 is approximate in that small radial velocities arising from the effect of viscosity variation on the flow field are neglected, so that terms like $v_r \, \partial T/\partial r$ do not appear. In addition, as is usual, it is assumed that axial conduction of heat is unimportant, so that the term $\partial(k \, \partial T/\partial z)/\partial z$ does not appear. The fluid is assumed incompressible, and ρ is taken as a constant. These assumptions will be valid for moderate temperature variations.

If the velocity profile is assumed to be perturbed by the varying temperature field, then the energy equation must be solved in conjunction with the dynamic equation for this flow, given by

$$0 = -\frac{\partial p}{\partial z} + \frac{1}{r}\frac{\partial}{\partial r}(r\tau) \qquad (2.67)$$

subject to the boundary conditions

$$v_z = 0 \quad \text{at} \quad r = R$$

$$\frac{\partial v_z}{\partial r} = 0 \quad \text{at} \quad r = 0$$

$$p = p_1 \quad \text{at} \quad z = 0$$

where, again, radial velocities are ignored.

If a model relating τ to $\dot{\gamma}$ is introduced, and if a functional form for the temperature dependence of the parameters of this model is introduced, the resulting set of equations constitutes a non-linear coupled boundary value problem which defies analytical solution. Indeed, so far, the problem has been solved only for cases which are of little value to the point of discussion here.

For example, if the fluid properties are assumed constant, in which case the isothermal velocity profile is known, solutions exist from which the temperature profile may be calculated as a function of axial position. Bird (8) gives solutions for the case of the power law fluid, with either the isothermal or adiabatic wall. Toor (9) solves a similar problem but allows for compressibility of the fluid and its accompanying heating or cooling effect. Siegel et al. (10) gives solutions for the newtonian fluid with boundary conditions specifying constant prescribed flux at the tube wall. The simplest case studied was that of Brinkman (11) for the newtonian fluid with temperature-independent properties, with either isothermal or adiabatic wall.

None of the studies named accounts for the alteration of the flow field arising from viscosity variation across the radius of the capillary, and so none of these studies may be used to correct viscometry data. The major use of the results of these workers is the *estimation* of the temperature rise experienced by the fluid. If the estimate of the maximum temperature rise is small, say less than 1°C., then one might judge that the results need not be corrected at all for heating effects. If the estimated temperature rise is large, then one is faced with the problem of rejecting the data or accepting the results as subject to significant error.

Experimental results (12) indicate that the wall of a capillary behaves in a nearly adiabatic manner under usual operating conditions. Hence estimates of temperature rise based upon the adiabatic theories are useful. Figure 2.8 shows Bird's solutions for the power law fluid, plotted as a dimensionless wall temperature as a function of a dimensionless axial variable, with n as a parameter. The dimensionless variables are defined as

$$\Theta = (T - T_0) \frac{4kK^{1/n}}{D^2 \tau_w^{(n+1)/n}} \left(\frac{3n+1}{2n}\right)^2 \tag{2.68}$$

and

$$Z = \frac{4}{\text{Pé}} \frac{n+1}{3n+1} \frac{z}{D} \tag{2.69}$$

τ_w is just the wall shear stress ($\tau_w = (D \Delta P/4L)$), and Pé is the Péclet number (Pé $= D\langle V \rangle \rho \hat{C}_p / k$).

As an illustration of the use of these equations, consider a piece of data taken from Table 2.1.

$D = 0.271$ cm.
$\langle V \rangle = 400$ cm./sec.
$\tau_w = 4250$ dynes/cm.^{2}
$K = 97$ (dynes/cm.^{2}) sec.$^{0.4}$
$n = 0.4$

$L = 94$ cm.
$\rho = 1.0$ g./cm.^{3}
$\hat{C}_P = 1.0$ cal./g.-°C.
$k = 1.4 \ 10^{-3}$ cal./sec.-cm.-°C.

assumed to be the same as the values for the solvent, water

At the tube exit, $z = L$, one finds $Z = 10^{-2}$. From Fig. 2.8 and Eq. 2.68 $T - T_0$ is calculated to be about 2°C.

It is of interest to examine the manner in which variables such as $\langle V \rangle$ and D affect the maximum temperature rise. This can be seen by writing Eq. 2.68 as

$$T - T_0 = \frac{D^2 K}{4k} \dot{\gamma}^{n+1} \left(\frac{2n}{3n+1}\right)^2 \Theta\left(n, \frac{z}{D}, \text{Pé}\right) \tag{2.70}$$

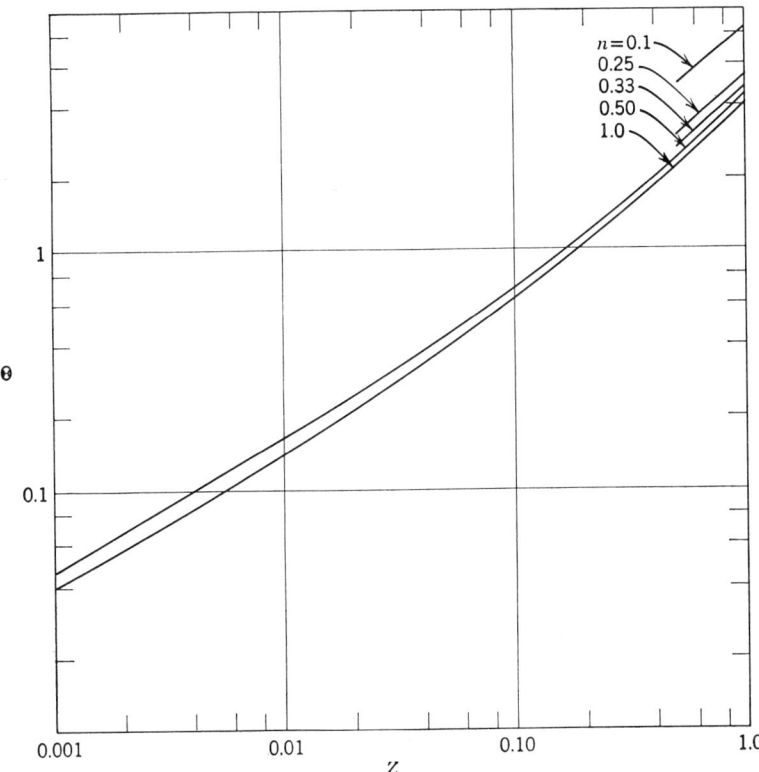

Fig. 2.8. Dimensionless wall temperature rise down the length of a capillary, due to viscous heating.

Nomograph for Estimation of Temperature Rise due to Viscous Dissipation

The nomograph on the facing page is based upon the solution for fully developed capillary flow of a power law fluid (8). That solution gives the temperature rise at the wall of the capillary, at some axial position z, under the assumption that the wall is adiabatic. To simplify the nomograph, approximations were made which essentially remove any dependence on the power law index n. In examples tested, the nomograph yields a temperature rise within a factor of 2 of the analytical solution. Hence, it provides a rapid estimate of the order of magnitude of the effect of viscous dissipation in a highly viscous capillary flow.

All scales are in cm.-g.-sec. units. Conversion factors are given below. A short table of thermal properties of common polymers is also given. Note, especially, that k must be in cm.-g.-sec. units.

Move from left to right across the nomograph. For a given set of data, begin by connecting points on the $\dot{\gamma}$ and τ scales, and find the intersection of this line with Reference scale 1. Connect that point, through D, to Reference scale 2, and so on across the other scales to the temperature scale. The value of ΔT is an estimate of the temperature rise at the wall of the capillary, at the given value of z/D.

Thermal Properties
(Numbers given are typical of various reported literature values)

	$k\left(\dfrac{\text{g.cm.}}{\text{sec.}^3{}^\circ\text{C.}}\right)$	$k/\rho Cp\left(\dfrac{\text{cm.}^2}{\text{sec.}}\right)$
Benzene	$1.5\ 10^4$	$8.5\ 10^{-4}$
Water	$6.3\ 10^4$	$1.5\ 10^{-3}$
Polybutadiene/styrene copolymer	$2.1\ 10^4$	$1.2\ 10^{-3}$
Polybutene	$1.2\ 10^4$	$7.4\ 10^{-4}$
Polyethylene	$3.3\ 10^4$	$1.3\ 10^{-3}$
Polyethylene terephthalate	$1.2\ 10^4$	$5.1\ 10^{-4}$
Polystyrene	$1.2\ 10^4$	$6.2\ 10^{-4}$

Conversion factors

To obtain k in g.-cm.-sec.$^{-3}$°C.$^{-1}$, multiply
   k in cal. sec.$^{-1}$ cm.$^{-1}$ °C.$^{-1}$ by 4.2×10^7
   k in BTU hr.$^{-1}$ ft.$^{-1}$ °F.$^{-1}$ by 1.7×10^5

To obtain $k/\rho Cp$ in cm.^{2} sec.$^{-1}$, multiply
   $k/\rho Cp$ in ft.^{2} hr.$^{-1}$ by 0.26

To obtain τ in g.-cm.$^{-1}$-sec.$^{-2}$ (dyne cm.$^{-2}$), multiply
   τ in lb.f ft.$^{-2}$ by 480
   τ in lb.f in.$^{-2}$ by 6.9×10^4

To obtain D in cm., multiply
   D in inches by 2.54
   D in mils by 2.54×10^{-3}

It should suffice to use nominal values for $\dot{\gamma}$ and τ, i.e., $\dot{\gamma} = 8\langle V\rangle/D$ and $\tau = D\,\Delta P/4L$. For polymer melts, $\dot{\gamma}$ would be underestimated but τ would be overestimated (assuming no corrections are made to ΔP) and the errors would tend to be compensatory.

EXPERIMENTAL TECHNIQUES 33

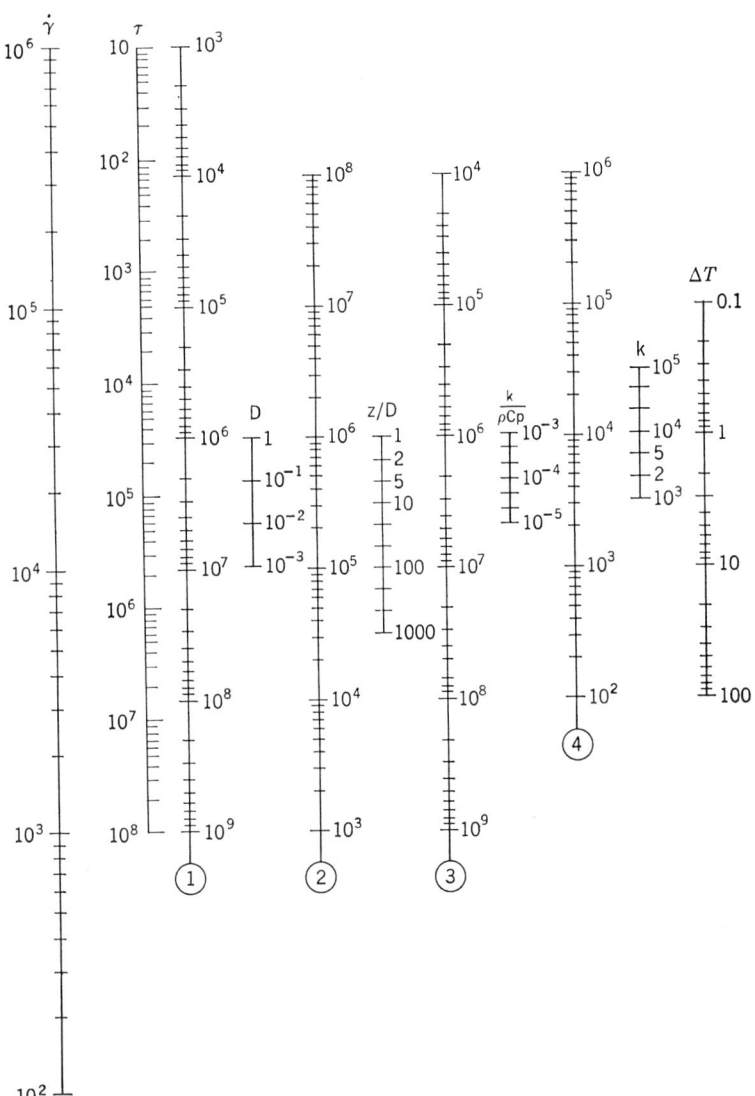

where τ_w has been replaced by $K\dot\gamma^n$. Assume now that one has a given fluid under study, and wishes to perform experiments at high shear rates. Assume also that

$$\dot\gamma = \frac{3n+1}{4n}\frac{8\langle V\rangle}{D}$$

and that a choice must be made between increasing $\dot\gamma$ by increasing $\langle V\rangle$ or by decreasing D. Suppose, for the sake of being definite, that $\dot\gamma$ is to be increased by a factor of 4. If this is done by decreasing D by a factor of 4, then the coefficient of Θ is *decreased* by a factor of $(\frac{1}{4})^2(4)^{n+1}$. For $n=1$ the factor is unchanged, while for $n=\frac{1}{2}$ the factor is halved. On the other hand, if $\dot\gamma$ is increased by increasing $\langle V\rangle$ by 4, at constant D, then the coefficient of Θ increases directly as $\dot\gamma^{n+1} = (4)^{n+1}$. For $n=1$ and $\frac{1}{2}$ this factor is 16 and 8, respectively. Clearly the heating effect is minimized if $\dot\gamma$ is increased by going to smaller capillaries. Actually Θ itself is a function of D and $\langle V\rangle$ through the Péclet number. However, the Péclet number dependence is weak under usual experimental conditions, and these conclusions hold. Of course, it has been assumed that K and n are unchanged by an increase in $\dot\gamma$. While this is not exactly true, in the sense that no fluid is truly power law in behavior, the approximation does not alter the conclusions drawn here to a significant extent.

Capillary viscometry of highly viscous materials should always be accompanied by an *estimate* of temperature rise due to viscous heating. It is true that many applications of viscosity data simply do not warrant the effort required to ascertain and correct for nonisothermal effects. On the other hand, one can find studies in the literature which draw conclusions based on the assumption that the high shear rate viscosity of highly viscous polymer melts is free of viscous heating perturbations.

In order to reduce the effort of temperature rise estimation, a nomograph has been prepared. Directions for its use are on page 32. This nomograph yields the wall temperature at the outlet of a capillary (or at any axial position in the capillary). It is based on Eq. 2.70 and Fig. 2.8, and neglects the effect of non-newtonian flow in that the value $n=\frac{1}{2}$ has been used in the approximations. In cases tested, the author has found it to be accurate to within a factor of 2, in comparison with the use of the exact relationships.

As an example of the importance of a temperature rise calculation, consider a study of viscometry of polydimethylsiloxane (12a). The authors cite, as an example, data taken at a shear stress of 10^7 dynes/cm.² (for which the shear rate is about 2×10^5 sec.$^{-1}$), in a die of diameter 0.50 mm. and L/D of 2. From the nomograph, the maximum *wall* temperature at the die exit would appear to be in excess of 20°C. The authors clearly recognized the importance of viscous heating, and performed an experiment in which a thermometer was placed in the stream of extrudate from the die. They measured only a 4–5°C. temperature rise, and carried out no viscosity correction for this temperature rise since it corresponded to only a 15% reduction in viscosity. It seems likely, however, that the thermometer measured the *bulk* temperature of the extrudate, rather than the wall temperature, which would be larger than the bulk temperature by a factor of 3 or 4. The significant point is that these authors would like to interpret some features of molecular flow mechanisms based on their observation of viscosity data at high shear rates. Their data are probably not significant in this region because of the high temperature rises encountered, but not recognized.

b. Couette Flow. If the Couette flow of Fig. 2.3 is considered and viscous dissipation is accounted for, then one must solve the dynamic and energy equations together. These equations would be

$$0 = \frac{d}{dr}(r^2\tau) \qquad (2.71)$$

$$0 = \frac{1}{r}\frac{d}{dr}\left(rk\frac{dT}{dr}\right) + \Phi \qquad (2.72)$$

subject to boundary conditions

$$v_\theta = 0 \quad \text{and} \quad T = T_0 \quad \text{at} \quad r = R_0$$
$$v_\theta = V = 2\pi\Omega R \quad \text{and} \quad \left.\begin{matrix} T = T_0 \\ \text{or} \quad \frac{\partial T}{\partial r} = 0 \end{matrix}\right\} \text{at} \quad r = R$$

The nature of this system is such that T and v_θ are functions *only* of radial position, and so are described by *ordinary* differential equations. Because of this it is possible to introduce some realistic complications into the analysis which cannot be tolerated in the capillary flow problem, described by *partial* differential equations.

The technique outlined here follows that used by Turian and Bird (13) for a similar problem. It is necessary to assume a constitutive equation and a temperature dependence of fluid properties. The power law is a useful model for the rheological behavior; the fluid properties are taken to be functions of temperature expressible in a power series about some reference temperature at which properties are assumed known. Thus τ is given by

$$\tau = K\dot{\gamma}^n \qquad (2.73)$$

and K is taken as

$$K = K_0[1 - K_1(T - T_0) + K_2(T - T_0)^2 - \cdots] \qquad (2.74)$$

For moderate temperature changes the thermal conductivity may be taken as constant. The power law index n is also taken to be independent of temperature. This is found to be a good approximation; to do otherwise leads to an unmanageable problem.

One now solves Eqs. 2.71 and 2.72 by a perturbation method. The details are quite similar to those outlined in Ref. 13 and the reader is referred to that paper. The major results are presented below, according to whether one assumes the isothermal inner wall or the adiabatic inner wall.

1. Isothermal wall
   a. Maximum temperature rise

$$\frac{\theta_{max}}{B_R} = \frac{(T - T_0)_{max}}{T_0 B_R} = \frac{n^2 C_0^{1+(1/n)}}{4}\left[1 - \frac{n}{2}f(s, n)\left(1 - \ln\frac{n}{2}f(s, n)\right)\right] \quad (2.75)$$

   b. Shear rate at the moving wall

$$\dot{\gamma}_R = \dot{\gamma}_0\left\{1 - \frac{\beta B_R s n^2 C_0^{1+(2/n)}}{16}[1 - s^{-4/n} - n(1 - s^{-2/n})f(s, n)]\right\} \quad (2.76)$$

2. Adiabatic wall
   a. Maximum temperature rise

$$\frac{\theta_{max}}{B_R} = \frac{n^2 C_0^{1+(1/n)}}{4}\left[1 - s^{-2/n} - \frac{2}{n}s^{-2/n}\ln s\right] \quad (2.77)$$

   b. Shear rate at the moving wall

$$\dot{\gamma}_R = \dot{\gamma}_0\left[1 - \frac{\beta B_R s n^2 C_0^{1+(2/n)}}{16}\left(1 - s^{-4/n} - \frac{4}{n}s^{-2/n}\ln s\right)\right] \quad (2.78)$$

where $f(s, n) = (1 - s^{-2/n})/\ln s$, $C_0 = s^{2-n}(\dot{\gamma}_0/\Omega)^n$, and $\dot{\gamma}_0$ is given by Eq. 2.48.

In the expressions above B_R is the Brinkman number

$$B_R = \frac{R^{1-n}K_0 V^{1+n}}{kT_0} \quad (2.79)$$

and $\beta = K_1 T_0$ is a measure of the temperature sensitivity of K. If βB_R is not small compared to unity then the solutions above are inaccurate. In that case more terms in the perturbation series are required.

The expressions for $\dot{\gamma}_R$ are in the form of a correction factor times the isothermal shear rate. In order to use the correction factor it is necessary to have at hand values of fluid properties such as β, K_0, and n. Since the viscometry experiment is supposed to measure K_0 and n these values are not known *a priori*. This suggests that correction for non-isothermal effects must proceed by trial and error. The data are treated, as in Section 2 (p. 19), as if free of heating effects, and K_0 and n are estimated. The maximum temperature may then be estimated and a second set of data obtained at this temperature level. From estimates of K_0 at two temperatures an estimate of β may be obtained. From β, K_0, and n an approximate correction factor is obtained and new values of β, K_0, and n are calculated. The procedure is repeated until satisfactory convergence is obtained.

c. Cone and Plate. The effect of viscous heat generation in the cone and plate system was determined by Turian and Bird (13) for the newtonian fluid with temperature-dependent properties, and by Turian (14) for the power law and Ellis model fluids with temperature-dependent properties. It was assumed that the cone angle was sufficiently small that the simple shear flow of Eq. 2.65 would hold if the flow were isothermal.

Turian's results for the power law fluid are presented here. The terms corresponding to a variable thermal conductivity have been dropped.

1. Isothermal cone
 a. Maximum temperature

$$\theta_{max} = \tfrac{1}{4} B_R - \frac{B_R^2 \beta}{24n}\left(\frac{n}{4} - \frac{1}{16}\right) \tag{2.80}$$

 b. Torque

$$\mathcal{T} = \mathcal{T}_0 \left[1 - \frac{B_R \beta}{20n}\right] \tag{2.81}$$

2. Adiabatic cone
 a. Maximum temperature

$$\theta_{max} = \tfrac{1}{2} B_R - \frac{B_R^2 \beta}{24n}(4n - 1) \tag{2.82}$$

 b. Torque

$$\mathcal{T} = \mathcal{T}_0 \left[1 - \frac{B_R \beta}{5n}\right] \tag{2.83}$$

The Brinkman number is given by

$$B_R = K\left(\frac{\Omega}{\psi_0}\right)^{n-1} \frac{R^2 \Omega^2}{kT_0} \tag{2.84}$$

and $\mathcal{T}_0$ is the torque which would be measured in the absence of viscous heating, given by

$$\mathcal{T}_0 = \frac{2\pi R^3 K}{3}\left(\frac{\Omega}{\psi_0}\right)^n \tag{2.85}$$

for the power law fluid.

An iteration scheme for correcting cone and plate data suggests itself. One may take data over a range of shear rates and thereby estimate values of K and n. θ_{max} may then be estimated, and a judgment made as to the need for correction. If it is decided to correct the data then an estimate of β is required. This may be obtained from a set of data taken at a bath temperature corresponding to T_{max}.

From these estimates the bracketed correction term in either Eq. 2.81 or 2.83 may be calculated. Since $\mathcal{T}$ is measured it is then possible to calculate $\mathcal{T}_0$. From Eq. 2.85 K and n may be calculated from two measurements of $\mathcal{T}$ at two different values of Ω/ψ_0. The correction term may then be improved and more accurate values of K and n may be estimated.

If the cone surface is isothermal then the value of K_0 is obtained. If the surface is adiabatic then K must be corrected to the bath temperature by using Eq. 2.74. Whether the cone is adiabatic or isothermal will depend upon the design of the viscometer. If the cone is not instrumented for internal cooling, then it will behave nearly adiabatically, once equilibrium is reached.

d. The Rate of Approach to Equilibrium. The question of the time required to reach thermal equilibrium turns out to be of some importance in planning the method of taking viscometric data. Many viscometers have no provision for maintaining the bounding surfaces at a fixed temperature. In some rotational viscometers, the cup can be immersed in a temperature bath, but the inner bob will slowly rise in temperature as viscous dissipation proceeds. Eventually, a thermal equilibrium will be achieved in which the bob behaves in an adiabatic manner, assuming that little heat transfer occurs through the bob shaft. In such a situation, if one waits long enough before taking an instrument reading, the viscosity data may be corrected for viscous heating using an analysis based upon the assumption that the cup is isothermal and the *bob adiabatic*.

It is possible that the thermal response of the bob is so slow, in comparison to the response of the fluid, that the bob surface may remain nearly isothermal while the temperature profile in the fluid develops to (nearly) its equilibrium condition. Under these circumstances, one might be able to take data shortly after shearing begins, and correct the data using the *isothermal bob* analysis. Alternatively, one might investigate the possibility that a reading could be taken so quickly that the *fluid* is still essentially isothermal and, hence, requires *no correction* for heating effects.

A theoretical study of these questions (15) indicates some guidelines for "strategy" in obtaining data under conditions subject to significant viscous heating. A mathematical model was set up for the temperature response of a newtonian fluid confined between infinite parallel plates. It was assumed that one plate was maintained at the initial (uniform) temperature of the fluid. The second (moving) plate was taken to be of finite thickness, and was allowed to respond to the temperature rise of the fluid as dissipation proceeded. The attempt was to model the thermal

Table 2.3

Sequence of Events in the Thermal Response of Couette Flow with a Conducting Boundary

Log time (sec.)		I Organic solution $k = 3.3 \times 10^{-4}$ (cal./sec.-cm.-°K.) $k/\rho\hat{C}_p = 10^{-3}$ (cm.²/sec.)		II Aqueous solution $k = 1.4 \times 10^{-3}$ (cal./sec.-cm.-°K.) $k/\rho\hat{C}_p = 6 \times 10^{-3}$ (cm.²/sec.)
−3				−3
	——— A	Velocity profile fully developed	A ———	
			B ———	
−2				−2
	——— B	Midplane temperature deviates from initial condition		
−1				−1
0				
		Midplane temperature reaches equilibrium	C ———	0
	——— C		D ———	
1				1
	——— D	Bob surface temperature deviates from initial condition		
2				2
	——— E	Bob surface temperature reaches equilibrium	E ———	
3				3

```
///////////////////////////////  ↑
////////////// Bob //////////////  c
///////////////////////////////  ↓
-------- Fluid --------           ↕ b
///////// Isothermal surface ////  ↑
```

For this example, $b = 0.1$ cm., $c = 1$ cm., the bob thermal diffusivity is 0.11 cm.²/sec., the bob thermal conductivity is 0.1 cal./sec.-cm.-°K., and the fluid kinematic viscosity is 1 poise.

response of a Couette flow in which the bob absorbed part of the heat generated by the shearing.

The sequence of events can be outlined with Table 2.3 as an example of results for two specific cases studied. The examples in columns I and II of Table 2.3 correspond to values typical of organic solutions or aqueous solutions, respectively. The velocity profile is seen to develop very rapidly. The midplane temperature of the fluid begins to rise and undergoes 5% of its ultimate change in a very short time. Hence, under practical experimental conditions, it would not be possible to obtain an instrument reading before the initial isothermal state is destroyed. The midplane temperature reaches 95% of its equilibrium change at times of the order of 1 sec. It is not until *later* times that the bob surface begins to deviate from its initial isothermal state. Hence, between times C and D, the temperature profile is in equilibrium with the isothermal bob boundary condition. The question, then, is whether one can obtain a viscometer reading in that time period. For the organic solution, a reading must be taken between 3 and 13 sec. after start-up, for the particular example illustrated here. This would not be particularly difficult. For the aqueous solution, a reading would have to be taken between 0.5 and 3 sec. The response time of the measuring system itself (such as the damping of a spring-dial system) might preclude such a measurement.*

After a considerable passage of time, the bob surface reaches equilibrium and the bob behaves in an adiabatic manner. One would have to wait 200 sec. for the aqueous solution and 840 sec. for the organic solution, for the example cited. In the case of the aqueous solution, then, one would be wise to wait about 3 min. after start-up, and correct the viscosity measurement with an *adiabatic bob* calculation. For the organic solution, if the instrument response permits, one would prefer to take a reading after about 10 sec., and correct with the *isothermal bob* theory, rather than wait nearly 15 min. for a single reading.

Of course, each specific case would have to be judged from a knowledge of the thermal properties of the fluid and bob under examination. Dimensionless solutions from which this information may be obtained are presented in Ref. 15. The point to be made here is that if one must correct viscometric data, the method whereby the data are obtained may seriously influence the degree to which the theory used for correction is applicable.

B. Normal Stress Measurement

To this point the discussion has centered about the techniques of "viscometry," the measurement of the shear stresses, or the corresponding

---

* Furthermore, a significantly elastic fluid would require more time for the development of the shearing stress than the newtonian example used here [see Eq. 3.135].

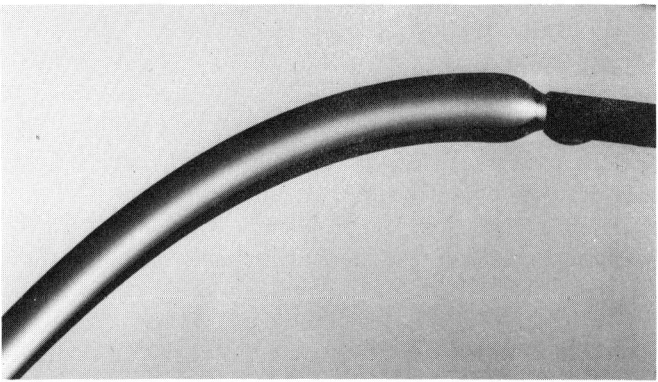

Fig. 2.9. Expansion of a viscoelastic fluid extruded from a long capillary. 8.1% solution of PIB L100 in toluene.

forces, required to maintain a simple shear flow. It is known, however, that many fluids, when subjected to a simple shear flow, develop not only shear stresses but also normal stresses. These normal stresses manifest themselves in the so-called "normal stress phenomena" which commonly occur in viscometric experiments, as well as in industrially important flows. For example, when a rod (such as the shaft of a stirrer) is rotated about its axis perpendicular to the free surface of a newtonian liquid, the liquid surface is depressed in the neighborhood of the rod, as a consequence of centrifugal forces which accompany the induced rotational flow. Some fluids, however, are observed to develop an *elevated* surface at the rod. This "rod climbing" effect is the best known normal stress phenomenon, and is generally called the "Weissenberg effect."

When certain non-newtonian fluids are ejected from an orifice or tube, the resulting jet is commonly observed to expand to a diameter much larger than its initial ejection diameter. Figure 2.9 shows a photograph of such an expanding jet. This effect is sometimes called the "Barus phenomenon,"* and is usually referred to industrially as "die swell," or "extrudate swelling."

This section is concerned with attempts to obtain information of fundamental rheological significance from the measurement of normal stress phenomena. In contrast to the measurement of shear stresses, the measurement of normal stresses includes the isotropic stress $p\boldsymbol{\delta}$ in addition to the dynamic stress $\boldsymbol{\tau}$. In order to generate information specifically about the dynamic stress components, it is necessary to "remove" the influence

---

* This is apparently a misnomer (16).

of p from the measurements. This is most commonly achieved by presenting results in terms of stress *differences* since, for example

$$T_{11} - T_{22} = -p + \tau_{11} - (-p + \tau_{22}) = \tau_{11} - \tau_{22}$$

and the isotropic component does not appear. This fact was anticipated in defining normal stress coefficients Ψ_{ij} on p. 10. When this is not possible it may be necessary to impose some assumptions upon p.

The fundamental meaning of the isotropic component p is still not clear for a viscoelastic material. While one commonly maintains the definition

$$p = p_m = -\tfrac{1}{3}(T_{11} + T_{22} + T_{33})$$

for such fluids, so that p is, by definition, "mean normal stress," it is not obvious that this *must* be true. Indeed, Williams (17) has discussed an alternate treatment for p which shows that useful results may be obtained without defining p as p_m, although the results do not preclude the possibility that $p = p_m$.

As in the case of viscometry, the flows studied approximate simple shear flows. In contrast to the techniques of viscometry, many of the methods used for normal stress measurement are incompletely developed, in the sense that the effects of perturbations which cause deviations from a simple shear flow, such as exit effects in capillary flows, are not fully understood, and methods for removing the influence of such effects from data are subject to assumptions not yet theoretically justified. Because the field of normal stress measurement is relatively young, an insufficient body of data exists upon which to base strong *a posteriori* justification for some of the assumptions made in treating normal stress data. Despite these difficulties it seems appropriate to present descriptions of the analyses most commonly used in treating normal stress measurements.

1. Capillary Jet

When a fluid issues from a tube as a free jet it exerts an axial thrust upon the tube. This thrust can be calculated as the difference between the flux of axial momentum leaving the tube and the axial stresses which exist in the fluid as it is ejected. Thus the thrust F is

$$F = \int_0^R \rho v_z^2 2\pi r \, dr - \int_0^R T_{zz} 2\pi r \, dr \qquad (2.86)$$

subject to the assumption that surface tension and gravity do not contribute significantly to the axial momentum.

It is possible to invert this relationship by a technique very similar to that used in developing the Weissenberg-Rabinowitsch-Mooney equation and thereby solve for T_{zz} at the tube wall. The resultant expression is (18)

$$(T_{zz})_R = \rho \langle V \rangle^2 \left[\frac{3n+1}{n} - 2 \int_0^1 \left(\frac{v_z}{\langle V \rangle} \right)^2 \frac{r}{R} d\left(\frac{r}{R}\right) \right]$$

$$- \frac{F}{\pi R^2} \left(1 + \frac{1}{2n} \frac{d \ln F}{d \ln 8 \langle V \rangle / D} \right) \quad (2.87)$$

Since the shear rate at the tube wall is known through the Weissenberg-Rabinowitsch-Mooney equation, it is possible to obtain T_{zz} and the corresponding shear rate. This result is subject to the assumption that the velocity and stress distributions at the tube exit correspond to those for fully developed flow. n is the slope of the $\log \tau$ vs. $\log \dot{\gamma}$ curve at the conditions of interest.

It is desirable to introduce the components of $\boldsymbol{\tau}$ in place of the components of $\mathbf{T}$, since constitutive equations predict only the behavior of the dynamic stress tensor $\boldsymbol{\tau}$. This is done through the use of the expression

$$(T_{zz})_R = -p(R, L) + (\tau_{zz})_R \quad (2.88)$$

Since $(T_{zz})_R$ is measured, using Eq. 2.87, at the exit $(Z = L)$, one must know the isotropic stress component $p(R, L)$ at the tube exit in order to find $(\tau_{zz})_R$.

It might be assumed that $p(R, L)$ is just atmospheric pressure at the tube exit, but examination of the dynamic equations shows this to be false. The radial component of the equations of motion gives

$$0 = -\frac{\partial p}{\partial r} + \frac{\partial \tau_{rr}}{\partial r} + \frac{\tau_{rr} - \tau_{\theta\theta}}{r} \quad (2.89)$$

Integration from $r = 0$ to some arbitrary r gives

$$p(r, Z) = p(0, z) + \tau_{rr} + \int_0^r \frac{\tau_{rr} - \tau_{\theta\theta}}{r} dr \quad (2.90)$$

subject to the reasonable assumption that $\tau_{rr} = 0$ along the center line of the tube, where the deformation rate vanishes. Hence it follows that

$$(T_{11})_R = -p(0, L) + (\tau_{11} - \tau_{22})_R - \int_0^R \frac{\tau_{22} - \tau_{33}}{r} dr \quad (2.91)$$

where the generalized notation of p. 9 has been introduced. In the absence of dynamic normal stresses it follows that the only isotropic stress would be ambient pressure p_0, and $p(0, L) = p(R, L) = p_0$ would be true.

The axial stress T_{11} may also be determined from a jet expansion experiment. Let the final radius and velocity of the jet be R_j and V_j and assume that the liquid is free of stress (relative to the ambient pressure) at this point. Then the flux of momentum across any position beyond the region where these final values are achieved is $\pi \rho R_j^2 V_j^2$. But, since momentum is conserved in the jet, this quantity must just equal the thrust exerted on the tube. Hence, in Eq. 2.87, one can replace F by $\pi \rho R_j^2 V_j^2$ and obtain, after some simple algebra

$$(T_{11})_R = \rho \langle V \rangle^2 \left[\frac{3n+1}{n} - 2 \int_0^1 \left(\frac{v}{\langle V \rangle} \right)^2 \frac{r}{R} d\left(\frac{r}{R} \right) \right.$$
$$\left. - \frac{1}{n\chi^2} \left(1 + n - \frac{d \ln \chi}{d \ln 8 \langle V \rangle / D} \right) \right] \quad (2.92)$$

where $\chi = R_j/R$. Equation 2.91, of course, is still valid regardless of whether one measures thrust or expansion.

An intermediate result which proves useful in a later discussion is possible in terms of the normal stress *averaged* over the cross section of the tube. The average axial stress is given by (19)

$$\langle T_{11} \rangle = \rho V^2 \left(\frac{3n+1}{2n+1} - \frac{1}{\chi^2} \right) \quad (2.93)$$

In terms of this quantity the stress $(T_{11})_R$ may be calculated from

$$(T_{11})_R = \frac{\langle T_{11} \rangle}{2n} \left(\frac{d \ln \langle T_{11} \rangle}{d \ln 8 \langle V \rangle / D} + 2n \right) \quad (2.94)$$

The stress T_{11} is associated with the normal stress in equilibrium with the fully developed shear flow within the capillary. It has been shown, however, that a viscous contribution to T_{11} exists which arises from the relaxation of the velocity profile subsequent to ejection of the liquid as a free jet (20). In newtonian fluids, for which there is no normal stress *except* that developed by profile relaxation, this viscous stress can be large enough to cause the jet to swell. For Reynolds numbers larger than about 200, however, the viscous contribution is of negligible importance.

In non-newtonian fluids it is not possible *a priori* to separate the viscous effect from the primary normal stress effect. It is usually assumed that the viscous stresses are negligible under the experimental conditions which are peculiar to a study of jets of polymeric fluids. The validity of this assumption can be checked *a posteriori* by plotting values of T_{11} obtained with different diameter tubes against shear rate. Since the viscous effect would not be a function of shear rate, but rather would be a function of a viscous parameter, such as a non-newtonian Reynolds number, data taken in different diameter tubes would fall on a single curve, when plotted as T_{11} vs. $\dot{\gamma}$, only if free of viscous effects.

It has been demonstrated experimentally and argued theoretically that at very high flow rates the ratio R_j/R approaches a limiting value which depends only upon the slope n of the log τ vs. log $\dot{\gamma}$ curve for the particular fluid. For the power law fluid this limiting value is given by

$$\chi_\infty = R_j/R = \sqrt{(2n + 1)/(3n - 1)} \qquad (2.95)$$

and lies in the range $0.866 \leq \chi < 1$ for $1 \geq n > 0$.

It is not difficult to see, upon examining Eq. 2.93, that when χ approaches χ_∞ the calculation of T_{11} becomes subject to considerable error arising from the subtraction of numbers large compared to their differences. Hence, Eq. 2.93 does not give accurate results at very high shear rates, where small errors in the measurement of χ are exaggerated by the subtraction process.

The same problem exists with the capillary thrust technique, and the superior technique will be the one which offers a more precise determination of either χ or F at high shear rates.

As in the case of viscosity determination through capillary flow measurements, it is necessary that the tube be sufficiently long that the normal stress distribution be fully developed. As yet, no well-established criterion, analogous to Eq. 2.13, exists for estimating the required entrance length for normal stress development. It is possible to offer some conjectures, however, which can be subjected to experimental investigation.

One can define, qualitatively, a characteristic relaxation time for a fluid as the time required for the stress in the fluid to adjust to a step change in shear rate. Figure 2.10 gives a possible quantitative definition for a relaxation time. One would expect that the normal stresses are in equilibrium with the shear field in the tube if the residence time $L/\langle V \rangle$ is larger than the relaxation time λ. Hence a criterion for fully developed normal stresses would be

$$\frac{L_n/\langle V \rangle}{\lambda} = k_n \quad \text{or} \quad \frac{L_n}{D} = k_n \frac{\langle V \rangle \lambda}{D} \qquad (2.96)$$

Of course k_n must be established through experiment, but one would expect it to be of the order of unity. Estimates of λ (discussed more

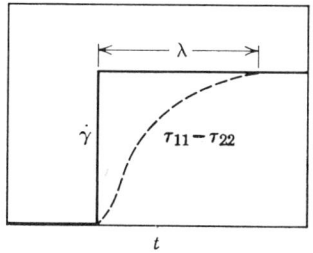

Fig. 2.10. Concept of a relaxation time, based on the hypothetical response of a viscoelastic fluid to a step change in shear rate.

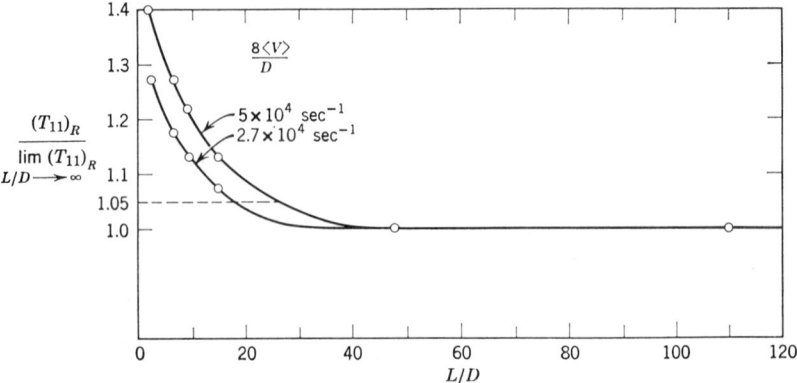

Fig. 2.11. Normal stress, based on jet expansion, as a function of capillary length. 3% solution of PIB L100 in tetralin at 25°C.

thoroughly in Chapter 4, p. 148) are of the order of magnitude of 10^{-2} sec. for fluids which show significant expansion. For typical values of $\langle V \rangle$ and D, say $\langle V \rangle = 500$ cm./sec. and $D = 0.1$ cm., a typical estimate for L_n/D is 50, consistent with experimental observations.

Entrance lengths for normal stresses are easily established experimentally (21). Figure 2.11 shows the axial stress $(T_{11})_R$, obtained from jet swelling measurements, plotted against L/D, for two values of shear rate. If the entrance length L_n is taken (arbitrarily) as that point beyond which $(T_{11})_R$ is within 5% of its value for very long tubes, then the results can be expressed in the form of Eq. 2.96 with $k_n = 3.3$. Bueche's relaxation time (p. 148) has been used for λ, with the viscosity average molecular weight assumed to be the appropriate average molecular weight (p. 151).

Since shear rates in capillary flows are generally high (around 10^4 sec.$^{-1}$), and since fluids which exhibit significant swelling are fairly viscous, heat generation within the capillary can be appreciable. In addition to producing a temperature different from the one which the fluid is presumed to have on the basis of the reservoir temperature, the temperature field alters the velocity profile. Since the velocity profile appears in Eqs. 2.87 and 2.92, it is clear that viscous heating effects could complicate the proper interpretation of normal stress measurement.

Recent experimental work by Gavis and Modan (22) has established a significant viscous heating effect on the expansion of newtonian jets. While these results are not directly applicable to non-newtonian fluids they serve to indicate that the "thermally perturbed" velocity profile affects the swelling only at Reynolds numbers of the order of 10 or less.

Such a low Reynolds number generally occurs at a velocity too low for execution of a successful jet experiment with viscoelastic fluids.

If the effects mentioned (profile relaxation, entrance stresses, and viscous heat generation) are unimportant, then Eq. 2.87 allows the calculation of $(T_{11})_R$ from appropriate data, and Eq. 2.91 then yields a relationship among the normal stress difference $(\tau_{11} - \tau_{22})$ the normal stress difference $(\tau_{22} - \tau_{33})$, and the isotropic stress $p(0, L)$. Since the shear rate vanishes along the jet center line, one would expect $p(0, L)$ to be free of any influence from the dynamic stress τ. It seems reasonable to suppose, then, that $p(0, L)$ is just the ambient pressure p_0. Furthermore, experimental results (23, 24) indicate that the stress difference $\tau_{22} - \tau_{33}$ is much smaller than the difference $\tau_{11} - \tau_{22}$. Hence, as a good approximation, the measured quantity $(T_{11})_R$ gives $(\tau_{11} - \tau_{22})_R$ directly. Since the shear rate at the wall is known from viscometry data

$$(\dot{\gamma}_R = [(3n + 1)/4n](8\langle V \rangle/D)),$$

one may determine this normal stress difference as a function of shear rate, or, equivalently, the variation of the material property Ψ_{12} with shear rate. Of course, the same remarks hold if $(T_{11})_R$ is determined, through Eq. 2.92, from jet swelling data.

Metzner and Shertzer (18) have presented experimental data using both the thrust and swelling techniques. Figure 2.12 shows their results, replotted as Ψ_{12} vs. $\dot{\gamma}$. The good agreement between the two methods indicates the reliability of either technique for the determination of $(T_{11})_R$. However, since the axial stress measured by either thrust or swelling would be perturbed in the same way by deviations from simple shear flow at the tube exit, the results do not allow one to reject the possibility of significant effects of profile relaxation, for example. The good agreement among the data taken at different tube diameters, however, lends support to the belief that the stress measured is free of such perturbations.

What appears to be a lower branch on the Ψ_{12} vs. $\dot{\gamma}$ curve in Fig. 2.12 is really the result of experimental uncertainty. The thrust data on this branch correspond to the lowest thrusts measured of all the data shown. The "expansion" data on this branch actually involve jet *contraction* of a magnitude close to that given by the inertial limit $[(2n + 1)/(3n + 1)]^{1/2}$. Hence these data are inaccurate but they are included to illustrate the experimental problems one faces in performing experiments of this type.

2. Other Capillary Measurements

Under usual conditions the jet experiments involve shear rates in the range 10^4–10^5 sec.$^{-1}$. To extend the range over which useful normal stress

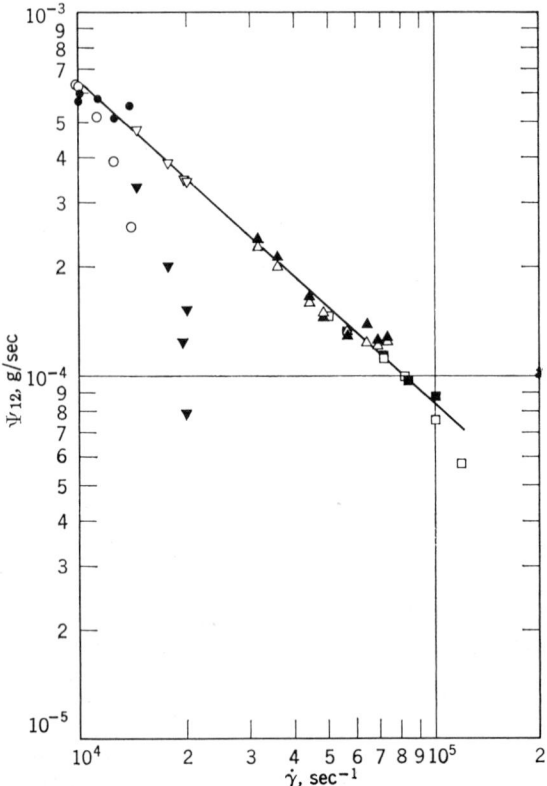

Fig. 2.12. Comparison of jet expansion with jet thrust data, based on Shertzer and Metzner (18). Filled symbols are jet expansion data.

data may be obtained in capillary flow, Savins (25) suggested a pitot tube method suitable to lower shear rates. With reference to Fig. 2.13, Savins' method would measure the pressures P_i and P_w. P_i would be an "impact" pressure, involving the axial normal stress in addition to the dynamic pressure $\frac{1}{2}\rho v_1^2(r)$, or

$$P_i = -T_{11}(r) + \tfrac{1}{2}\rho v_1^2(r) \tag{2.97}$$

The wall pressure P_w would be just $-T_{22}(R)$.

From the dynamic equations for laminar Poiseuille flow, one finds, for the radial direction

$$T_{22}(R) = -p(0, z) - \int_0^R (\tau_{22} - \tau_{33}) \frac{dr}{r} \tag{2.98}$$

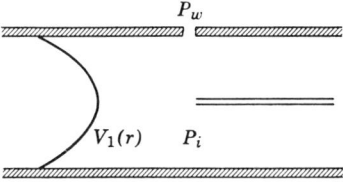

Fig. 2.13. Definition sketch for normal stress measurement using a pitot tube in laminar Poiseuille flow.

and, for the axial direction, Eq. 2.91, with R and L replaced by r and z

$$T_{11}(r) = -p(0, z) + (\tau_{11} - \tau_{22}) - \int_0^r (\tau_{22} - \tau_{33}) \frac{dr}{r} \quad (2.99)$$

One may then express the pressure difference $P_i - P_w$ as

$$P_i - P_w = -(\tau_{11} - \tau_{22}) + \tfrac{1}{2}\rho v_1^2(r) - \int_r^R (\tau_{22} - \tau_{33}) \frac{dr}{r} \quad (2.100)$$

As noted earlier, experimental evidence (23, 24) tends to indicate that the normal stress difference $\tau_{22} - \tau_{33}$ is considerably smaller than the normal stress difference $\tau_{11} - \tau_{22}$. Hence, as an approximation, the pressure measurement $P_i - P_w$ allows the determination of the normal stress difference $\tau_{11} - \tau_{22}$, assuming that the velocity profile is known. If viscous heating is significant the velocity profile will be perturbed, and the calculation of $\tau_{11} - \tau_{22}$ will be in error. At the low shear rates for which this method is proposed viscous heating effects are likely to be minimal. A thermocouple attached to the pitot tube would serve to indicate the presence of a significant temperature variation across the capillary.

One should also note, as Savins points out, that Eq. 2.100 indicates that a pitot tube may not be used directly to measure the velocity profile in a fluid which exhibits normal stress phenomena.

Any pitot tube measurement is subject to the criticism that the probe distorts the flow, and hence does not measure the velocity profile which would exist if the probe were not present. The success of pitot tube measurements in newtonian flows lends some confidence to its use in more complex materials. A dissenting view, however, is presented by Metzner and Astarita (26), who point out that a viscoelastic flow in the neighborhood of an obstruction may be considerably different from a newtonian flow which is kinematically similar at some distance from the obstruction. The difference arises primarily from a markedly different behavior of the boundary layer in viscoelastic stagnation flows. Evidence, based on markedly reduced heat transfer coefficients from heated cylinders placed in a stream of a viscoelastic fluid, supports this view.

According to the discussion of Metzner and Astarita, the pitot method proposed by Savins would be subject to significant error in highly elastic fluids at high shear rates. While their analysis seems to support a reasonable objection to the pitot tube method, only experimental evidence will be able to settle the point with any useful degree of certainty.

A motion closely related to the capillary flow is the axial flow of material confined to the annular space between concentric cylinders. A particular advantage of this configuration is that radial pressure measurements lead to evaluation of the normal stress difference $\tau_{22} - \tau_{33}$. This is most easily seen by taking the radial component of the dynamic equations and performing an integration in the radial direction. The result is

$$\Delta P_R = T_{22}(R) - T_{22}(sR) = -\int_{sR}^{R} (\tau_{22} - \tau_{33}) \frac{dr}{r} \quad (2.101)$$

where R and sR are the radii of the outer and inner cylinders, respectively. The pressure difference ΔP_R is just the quantity which would be measured by two taps placed opposite each other in the surface of the walls of the annulus.

Huppler (23) and Hayes and Tanner (27) present experimental results obtained with polymer solutions in such an axial annular flow. The form of Eq. 2.101 makes it necessary to assume a model for $\tau_{22} - \tau_{33}$ as a function of shear rate. The parameters of the model are then obtained by curve-fitting the model to the data. Huppler's results show clearly that $\tau_{22} - \tau_{33}$ is small compared to $\tau_{11} - \tau_{22}$ for a number of dilute polymer solutions.

3. Normal Stresses in the Cone and Plate System

In discussing the cone and plate system as a viscometer it was stated that, with good approximation, the velocity field is

$$v_\phi = v_1 = r\Omega\psi/\psi_0$$

and the shear rate is $\dot{\gamma} = \Omega/\psi_0$. The r-component of the dynamic equations, in spherical coordinates, is reducible to the form

$$-\rho\frac{v_1^2}{r} = -\frac{\partial p}{\partial r} + \frac{\partial \tau_{33}}{\partial r} - \frac{\tau_{11} + \tau_{22} - 2\tau_{33}}{r} \quad (2.102)$$

where the general coordinate notation (Table 1.6) has been used. Since shear rate is independent of r, and since τ is a function only of shear rate, terms such as $\partial \tau_{33}/\partial r$ are identically zero. The stress measured by pressure taps in the plate ($\psi = 0$) is just

$$T_{22}(r, 0) = -p(r, 0) + \tau_{22} \quad (2.103)$$

EXPERIMENTAL TECHNIQUES 51

If T_{22} is introduced into the dynamic equation, the result is

$$\frac{\partial T_{22}}{\partial r} = -\left(\frac{\rho v_1^2}{r} - \frac{\chi(\dot{\gamma})}{r}\right) \tag{2.104}$$

where $\chi = \tau_{11} + \tau_{22} - 2\tau_{33}$. If the isotropic component of **T** is taken as the mean normal stress, then $\chi = -3\tau_{33}$ (*cf.* p. 3). While most investigators have made this identification, Williams (17), for example, chooses not to (*cf.* p. 42) and so works with the function χ. If Eq. 2.104 holds at the plate surface ($\psi = 0$), then the measured stress distribution should obey

$$\frac{\partial T_{22}}{\partial r} = -\frac{3\tau_{33}}{r} \quad \text{or} \quad \frac{\partial T_{22}}{\partial \ln r} = -3\tau_{33} = \text{constant} \tag{2.105}$$

Deviations between this prediction and experimental results may be ascribed to inertial effects which cause the velocity profile to deviate from the simple shear expression assumed above.

A first-order estimate of inertial effects is possible for the newtonian fluid if the simple shear velocity profile is inserted in Eq. 2.102, so that

$$\frac{1}{\rho}\frac{\partial p}{\partial r} = \frac{v_1^2}{r} = r\Omega^2(\psi/\psi_0)^2 \tag{2.106}$$

Integration yields

$$(p - p_0) = \tfrac{1}{2}\rho\Omega^2(\psi/\psi_0)^2 r^2 + \rho C_N(\psi) \tag{2.107}$$

where p_0 is some arbitrary pressure and C_N is a function of ψ but not r. For $\psi = 0$, the pressure along the plate is independent of radial position, according to this result.

Experimental measurements are not in agreement with this prediction; instead they indicate a variation of p with r. A possible interpretation of this result seems to have no physical foundation, but does lead to a prediction consistent with experiment. It is claimed that the stress measured at the plate is really the ψ-average of the stress distribution between the cone and plate. The measured stress is thus thought to be equal to

$$\bar{p} = \frac{1}{\psi_0}\int_0^{\psi_0} p\, d\psi = p_0 + \tfrac{1}{6}\rho\Omega^2 r^2 + \rho\bar{C}_N \tag{2.108}$$

Williams (17) shows that the condition of incompressibility of the fluid demands that $\bar{C}_N = -\tfrac{1}{10}\Omega^2 R^2$, so that, finally, for a newtonian fluid

$$(\bar{p} - p_0) = -(\bar{T}_{22} + p_0) = \frac{1}{6}\rho\Omega R^2\left(\frac{r^2}{R^2} - \frac{3}{5}\right) \tag{2.109}$$

A detailed evaluation of the dynamic equations for a non-newtonian fluid, coupled with the assumption that the stress measured along the plate surface is

$$\bar{T}_{22} = \frac{1}{\psi_0} \int_0^{\psi_0} (-p + \tau_{22}) \, d\psi$$

yields the result

$$\bar{T}_{22} + p_0 - \tau_{22} = -3\tau_{33}\left(\frac{1}{3} - \ln\frac{R}{r}\right) - \frac{1}{6}\rho\Omega^2 R^2\left(\frac{r^2}{R^2} - \frac{3}{5}\right) \quad (2.110)$$

τ_{22} is not averaged because it is a function only of shear rate and so is independent of ψ. p_0 is taken to be the ambient pressure.

A plot of the measured quantity

$$\bar{T}_{22} + p_0 + \frac{1}{6}\rho\Omega^2 R^2\left(\frac{r^2}{R^2} - \frac{3}{5}\right)$$

against $\ln(R/r)$ should be linear, with a slope equal to $3\tau_{33}$ and a zero intercept at a radial position r_0 such that

$$\ln(R/r_0) = \tfrac{1}{3}(1 - \tau_{22}/\tau_{33}) \quad (2.111)$$

Hence both τ_{33} and τ_{22} may be obtained directly from such an experiment, if the inertial correction is known. Williams indicates that while no firm experimental confirmation of the inertial term is available, use of Eq. 2.110 gives consistent values for r_0 for various polymer solutions. Thus it is probably a good approximation to the inertial stresses set up at low shear rates in a system with a small cone angle (less than 5 deg.).

The inertial correction may be determined experimentally by measuring the stress distribution along the plate for a newtonian fluid. If inertial effects perturb the stress field without causing significant deviation of the velocity field from that assumed in Eq. 2.106, then a non-newtonian fluid will behave, with respect to inertial forces, like a newtonian fluid, since the shear field will be uniform throughout the fluid. Thus the inertial correction at a particular shear rate can be obtained from the measured stress distribution of a newtonian fluid whose viscosity is the same as the apparent viscosity of the non-newtonian fluid at that shear rate.

Since the inertial stresses depend upon Ω, but not upon ψ_0, data taken with different cone angles should be a function only of Ω/ψ_0 when the proper inertial corrections have been made. Thus one should always take data with more than a single cone angle whenever possible.

It should be noted that data exist (28, 29) which indicate that inertial effects can induce secondary flows in a cone and plate system. While this appears to occur only at large cone angles (above 5 deg.) for the rotational speeds usually employed, it is clear that such a deviation from a simple shear flow completely destroys the utility of the cone and plate system for the measurement of viscosity or normal stresses.

If the total vertical force exerted against the cone is measured, it is possible to determine the normal stress difference $\tau_{11} - \tau_{22}$ directly, without a measurement of the stress distribution in the system. For the small cone angles usually employed, this force may be calculated from

$$F = -2\pi \int_0^R T_{22} r \, dr \tag{2.112}$$

Integration by parts leads to

$$F = -2\pi \left[\tfrac{1}{2} R^2 T_{22}(R) - \int_0^R \tfrac{1}{2} r^2 \frac{\partial T_{22}}{\partial r} \, dr \right] \tag{2.113}$$

If Eq. 2.105 is used the result is

$$F = -2\pi [\tfrac{1}{2} R^2 T_{22}(R) + \tfrac{3}{2}\tau_{33} \tfrac{1}{2} R^2] \tag{2.114}$$

Finally, by replacing $T_{22}(R)$ by $-p(R) + \tau_{22}$, and by writing $\tfrac{3}{2}\tau_{33}$ as $\tau_{33} - \tfrac{1}{2}(\tau_{11} + \tau_{22})$, the bracketed quantity can be rearranged to the form

$$F = -\pi R^2 [(-p(R) + \tau_{33}) + \tfrac{1}{2}(\tau_{22} - \tau_{11})] \tag{2.115}$$

If the system is in equilibrium with the atmosphere on its outer boundary, then

$$-p(R) + \tau_{33} = T_{33}(R) = 0$$

and it follows that

$$\tau_{11} - \tau_{22} = 2F/\pi R^2 \tag{2.116}$$

Inertial effects may be introduced as before. The most common procedure is to correct F on the basis of experiments with newtonian fluids of appropriate viscosities. Then $\tau_{11} - \tau_{22}$ is calculated from F', a corrected force from which the inertial effect has been removed.

Some experimental results obtained by Shertzer and Metzner (18) are shown in Fig. 2.14. The total force method was used to calculate $\tau_{11} - \tau_{22}$, from which Ψ_{12} was determined. Data obtained with different cone angles appear to be consistent. The line at high shear rates represents the jet data of Fig. 2.12. While interpolation over a full decade of shear rate must be considered dangerous, the consistency of these two sets of results is very encouraging. The method proposed by Savins (p. 48) would be appropriate for filling in the gap in these data, if a reliable technique can be established for using a pitot tube in a viscoelastic fluid.

Results in the cone and plate system are very sensitive to small errors in alignment of the cone axis, and in position of the cone tip. Anyone seriously interested in working with a cone and plate instrument would be wise to read the thesis of Williams (30), which discusses in detail some of the techniques and precautions required to produce good data.

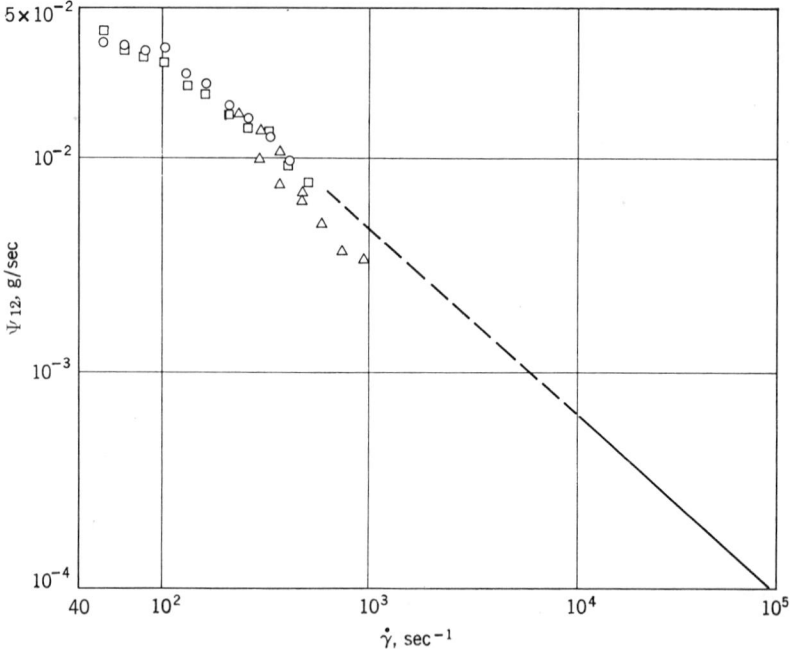

Fig. 2.14. Normal stress coefficient Ψ_{12} calculated from the thrust in cone and plate shearing (18). The solid line at high shear rates is from Fig. 2.12.

4. The Parallel Plate System

Another system useful for the measurement of normal stresses at low shear rates is the parallel plate torsional configuration, most easily achieved in the space between the bottoms of two concentric cups. Figure 2.15 shows the geometry to be considered.

If inertial effects are unimportant, and if edge effects at the radial boundary can be ignored, then the velocity profile may be shown to be (31)

$$v_1 = \Omega rz/L \tag{2.117}$$

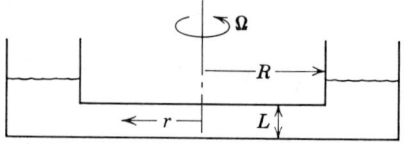

Fig. 2.15. Definition sketch for parallel plate torsional shear flow.

It follows, from Table 1.4, that the only non-zero components of the rate of deformation tensor are

$$\dot{\gamma} = \Delta_{12} = \Delta_{21} = \Omega r/L \qquad (2.118)$$

Note that the shear rate varies radially in this system.

In cylindrical coordinates the radial component of the dynamic equations is

$$0 = -\frac{\partial p}{\partial r} + \frac{1}{r}\frac{\partial}{\partial r}(r\tau_{33}) - \frac{\tau_{11}}{r} \qquad (2.119)$$

subject to the assumption that the inertial term $\rho v_1^2/r$ is small. (In this geometry the 3-coordinate is r, 2 is z, and 1 is θ.)

Since the axial stress T_{22} is the quantity usually measured in this system it is convenient to replace p by $\tau_{22} - T_{22}$, whence it follows that

$$\frac{dT_{22}}{dr} = \frac{d}{dr}(\tau_{22} - \tau_{33}) + \frac{1}{r}(\tau_{11} - \tau_{33}) \qquad (2.120)$$

Integration between the axis of rotation and some arbitrary radial position gives

$$T_{22}(r) + p(0) = \tau_{22} - \tau_{33} + \int_0^r \frac{\tau_{11} - \tau_{33}}{r} dr \qquad (2.121)$$

subject to the reasonable assumption that τ_{22} and τ_{33} are zero at $r = 0$, since the shear rate vanishes at that point.

Since T_{22} may be measured as a function of r, the right-hand sides of either Eq. 2.120 or Eq. 2.121 may be obtained. However, further information about the individual stress differences requires arbitrary assumptions about their relative values or their dependencies on shear rate.

A more useful result is obtained from the determination of the total vertical force exerted against the upper plate. The force F is given by

$$F = -2\pi \int_0^R T_{22}(r) r \, dr \qquad (2.122)$$

If $\dot{\gamma}$ is substituted for r through Eq. 2.118 and F is differentiated with respect to $\dot{\gamma}_R$, the result, after some algebraic manipulation and use of the Leibniz rule, is found to be (32)

$$-3\tau_{22}(\dot{\gamma}_R) = \frac{2F}{\pi R^2}\left(1 + \frac{1}{2}\frac{d\ln F}{d\ln \dot{\gamma}_R}\right) \qquad (2.123)$$

where $\dot{\gamma}_R = \Omega R/L$.

As in the case of the cone and plate system, F must be replaced by F', where F' includes a correction for inertial effects. It would appear that inertial stresses are far more serious in the parallel plate system than in the cone and plate, at comparable shear rates. Because the shear rate varies spatially in the parallel plate system, the inertial correction cannot be obtained exactly from experiments with a newtonian fluid. However, the newtonian stress estimate will considerably reduce the error due to inertial effects. Any residual anomalous behavior can be detected by plotting τ_{22} against the plate separation L, at constant shear rate. If the curve is not flat, it is suggested (24) that the τ_{22} curve be extrapolated to zero plate separation, and that value of τ_{22} used as a corrected stress.

Ginn and Metzner (24) present data shown in Fig. 2.16. The cone and plate results were obtained using Eq. 2.116 and so represent $\tau_{11} - \tau_{22}$. The parallel plate results follow from Eq. 2.123, and so give $-3\tau_{22}$. These two quantities are equal if, and only if, $\tau_{22} = \tau_{33}$. This equality, known as the Weissenberg relation, is predicted by many constitutive equations (33). Ginn and Metzner remark that the difference between the two quantities plotted in Fig. 2.16 is statistically significant at shear rates below 100 sec.$^{-1}$, and so contradicts the Weissenberg relation. It seems apparent, however, that the Weissenberg relation is a good approximation. This is

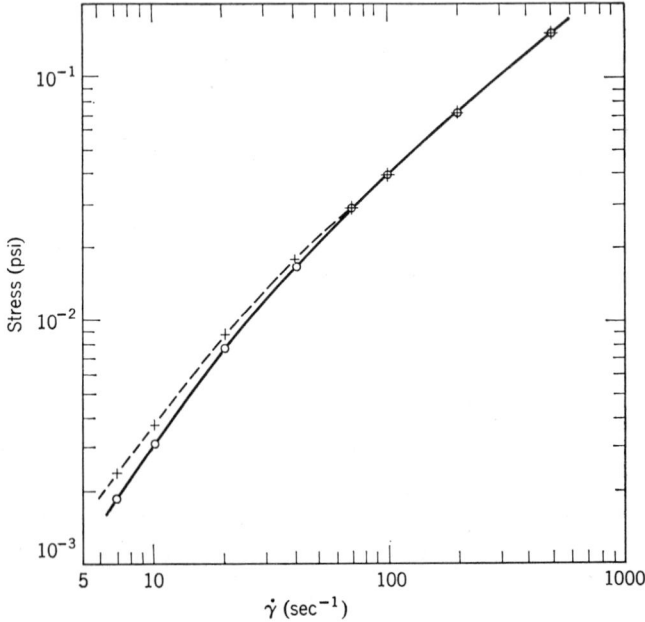

Fig. 2.16. Comparison of cone and plate data with parallel plate data for normal stress determination (24). Coincidence of data for the two systems requires that $\tau_{22} = \tau_{33}$.

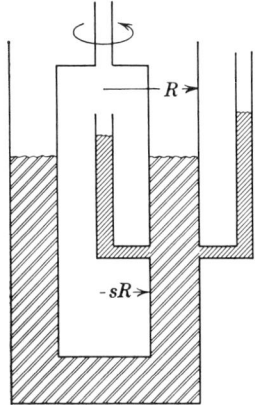

Fig. 2.17. Definition sketch for normal stress determination in cylindrical Couette flow.

gratifying in view of the simplification which often attends the neglect of the term $\tau_{22} - \tau_{33}$ in equations such as 2.100 and 2.121.

A serious error in the pressure measurements can be introduced by misalignment of the system. The plates must be parallel within very exacting limits. The reader contemplating experimental work would do well to read some detailed descriptions of the experience of other workers (24, 31).

5. A Concentric Cylinder (Couette) System

Padden and DeWitt (34) suggested and used a concentric cylinder apparatus to measure normal stresses. The primary measurement is the difference in radial stresses exerted at the two cylindrical surfaces. As illustrated in Fig. 2.17, this measurement is usually made by manometers.

If $_1$, $_2$, and $_3$ denote, respectively, the angular, radial, and axial directions, then the radial dynamic equation may be written as

$$-\rho \frac{v_1^2}{r} = -\frac{\partial p}{\partial r} + \frac{\partial \tau_{22}}{\partial r} + \frac{\tau_{22} - \tau_{11}}{r} \qquad (2.124)$$

subject to the usual assumption of a laminar simple shear flow. If the inertial term is neglected, Eq. 2.124 may be integrated between the inner and outer radii to give

$$\Delta(p - \tau_{22}) = -\Delta T_{22} = \int_{sR}^{R} \frac{\tau_{22} - \tau_{11}}{r} dr \qquad (2.125)^*$$

* The inertial term is $\mathscr{I} = \int_{sR}^{R} (\rho v_1^2/r) \, dr$, and may be estimated if the velocity field is known. If not neglected, $\mathscr{I}$ should be added to the right-hand side of Eq. 2.125.

Since $r^2\tau = $ constant (*cf.* p. 20), τ may be substituted for the variable of integration, and the resulting equation may be differentiated with respect to the shear rate at the inner wall $\dot{\gamma}_i$. This leads to

$$\frac{d\,\Delta T_{22}}{d\dot{\gamma}_i} = \frac{1}{2}\frac{d}{d\dot{\gamma}_i}\int_{\tau_i}^{\tau_0}\frac{\tau_{22}-\tau_{11}}{\tau}\,d\tau \qquad (2.126)$$

The differentiation is performed using the Leibniz rule with the result

$$\frac{d\,\Delta T_{22}}{d\dot{\gamma}_i} = \frac{1}{2}\left[\left(\frac{\tau_{22}-\tau_{11}}{\tau}\right)_0\frac{d\tau_0}{d\dot{\gamma}_i} - \left(\frac{\tau_{22}-\tau_{11}}{\tau}\right)_i\frac{d\tau_i}{d\dot{\gamma}_i}\right] \qquad (2.127)$$

where quantities subscripted 0 and i are to be evaluated at the conditions of the outer and inner walls, respectively.

For fixed geometry, since $\tau_0 = s^2\tau_i$, it follows that

$$\frac{d\tau_0}{d\dot{\gamma}_i} = s^2\frac{d\tau_i}{d\dot{\gamma}_i} \quad \text{or} \quad \frac{d\ln\tau_0}{d\ln\dot{\gamma}_i} = \frac{d\ln\tau_i}{d\ln\dot{\gamma}_i} = n \qquad (2.128)$$

where n, as defined here, is the slope of the double logarithmic plot of τ vs. $\dot{\gamma}$, and is presumed known as a function of $\dot{\gamma}$ from viscometric data for $\tau(\dot{\gamma})$.

Equation 2.127 may thus be rearranged to the form

$$\frac{2}{n}\frac{d\,\Delta T_{22}}{d\ln\dot{\gamma}_i} = (\tau_{22}-\tau_{11})_0 - (\tau_{22}-\tau_{11})_i \qquad (2.129)$$

The left-hand side of this equation may be measured as a function of shear rate $\dot{\gamma}_i$, with $\dot{\gamma}_i$ calculated from Eq. 2.45. As s approaches unity the shear rates at the inner and outer walls approach each other, and the difference on the right-hand side of Eq. 2.129 becomes subject to considerable error.

A useful approximation is possible if s is not close to unity. Let $\tau_0 = s^2\tau_i$, and suppose the viscosity behavior is approximated by the power law, $\tau = K\dot{\gamma}^n$. Then

$$\dot{\gamma}_0 = s^{2/n}\dot{\gamma}_i \qquad (2.130)$$

As a further approximation, probably valid for many fluids in the range of shear rates attained in this instrument (*cf.* p. 158)

$$(\tau_{22}-\tau_{11}) = A\dot{\gamma}^{2n} \qquad (2.131)$$

Then $(\tau_{22}-\tau_{11})_0 = s^4(\tau_{22}-\tau_{11})_i$ follows as an approximation. If $s = \frac{1}{2}$, the normal stress difference at the outer wall is only $\frac{1}{16}$ of the value at the inner wall. Hence, for $s < \frac{1}{2}$, a good approximation would be

$$\frac{2}{n}\frac{d\,\Delta T_{22}}{d\ln\dot{\gamma}_i} = -(\tau_{22}-\tau_{11})_i \qquad (2.132)$$

Of course, if these approximations are used, they could, and should, be verified by examination of the resultant data for $\tau_{22} - \tau_{11}$ vs. $\dot{\gamma}$. If these approximations are not valid, or if $s > \frac{1}{2}$, then the normal stress differences may be obtained from Eq. 2.129 by iteration, or by curve-fitting a model for $\tau_{22} - \tau_{11}$ as a function of $\dot{\gamma}$.

An iterative method has been given by Markovitz (35) which is very similar to the technique used in treating viscometric data in the coaxial cylinder viscometer. (See the analysis leading to Eq. 2.45.) His result may be written in the form

$$(\tau_{11} - \tau_{22})|_{\tau_i} = \sum_{p=0}^{\infty} \psi(s^{2p}\tau_i) \qquad (2.133)$$

where

$$\psi(\tau_i) = 2\tau_i \frac{d\Delta T_{22}}{d\tau_i}\bigg|_{\tau_i}$$

As before, τ_i is the shear stress at the inner cylinder. By plotting ΔT_{22} (with inertial correction if necessary) as a function of shear stress τ_i, values of ψ may be obtained, and therefrom values of $\tau_{11} - \tau_{22}$ may be determined as a function of shear stress. From viscometric data this may be converted, if desired, to a curve of $\tau_{11} - \tau_{22}$ as a function of shear rate.

In general, the series above must be truncated with some error resulting. If low enough stresses are achieved so that the normal stress functions reach a region of quadratic dependence on shear stress (as is the case in most polymer solutions), then Markovitz gives a closed-form solution for the series as

$$(\tau_{11} - \tau_{22})|_{\tau_i} = \sum_{p=0}^{P-1} \psi(s^{2p}\tau_i) + [\psi(s^{2P}\tau_i)/(1 - s^4)]$$

The number of terms in the summation is determined by the value of P such that $s^{2P}\tau_i$ is a shear stress in the quadratic region.

As in the parallel plate system, inertial stresses become important at high shear rates, and limit to some extent the range of shear rates which may be covered. Inertial stresses based on studies with newtonian fluids may be used to correct the data, but only approximately, since the shear rate varies across the annular gap significantly when s is not near unity. If the power law model is again assumed, for the sake of an approximation, it can be seen that the ratio of apparent viscosities $(\tau/\dot{\gamma})$ at the inner and outer cylinders is

$$\eta_i/\eta_0 = s^{2(1/n - 1)} \qquad (2.134)$$

If $s = \frac{1}{2}$ and $n = \frac{1}{2}$, then $\eta_i/\eta_0 = \frac{1}{4}$, and the actual non-newtonian inertial stresses might be poorly approximated by any newtonian estimate. With this precaution in mind, it may be noted that a first-order estimate of the inertial effect on ΔT_{22} is given by (36)

$$\mathscr{I} = \frac{\rho s^2 R^2}{8}(1 - s^4 + 4s^2 \ln s)\dot{\gamma}_i^2 \qquad (2.135)$$

where $\dot{\gamma}_i$ is the newtonian shear rate at the inner cylinder, given by Eq. 2.47. Within the spirit of this type of approximation it is possible to give an estimate for the inertial stress difference in a power law fluid, with the result

$$\mathscr{I} = \frac{\rho s^{2/n} R^2 n^2}{8} \left[s^{2/n}(1 - s^2) - \frac{2s^{2/n}(1 - s^{2(1-1/n)})}{1 - 1/n} + \frac{s^{2/n}(1 - s^{2(1-1/n)})}{1 - 2/n} \right] \dot{\gamma}_{in}^2 \qquad (2.136)$$

where $\dot{\gamma}_{in}$ is the power law shear rate, given by Eq. 2.48.

If $\mathscr{I}$ is estimated or measured, the foregoing equations are unchanged so long as ΔT_{22} is replaced by $\Delta T_{22} + \mathscr{I}$. (See Eq. 2.125 and footnote.) If inertial stresses are effectively removed by this method, then data for $\tau_{22} - \tau_{11}$ should fall on a single curve, when plotted against shear rate, with no dependence upon s or R. Markovitz (36) shows data taken at two different values of s. The data were corrected using Eq. 2.135. Except in the limit of very small shear rates the agreement between the two curves is rather poor. This probably indicates the failure of Eq. 2.135 to adequately account for inertial effects. A subsequent study, in which the inertial correction was made by calculating the *non-newtonian* velocity profile, gave excellent agreement for two values of s (35, 37).

C. Elongational Flows

In Chapter 1 simple elongation was defined as a flow with which another material property, the elongational viscosity, could be measured. One method of achieving a simple elongation is illustrated in Fig. 2.18. The upper end of a cylindrical sample is clamped and stationary. The lower end is clamped and moves with the velocity U. If end effects in the region of the clamps can be ignored then the deformation will be uniform, in the sense that the shrinking of the radius will not depend upon axial position, and the axial velocity will be uniform across the radius. Hence v_z is not a function of r, and v_r is not a function of z.

The axial velocity and radial velocity must satisfy the continuity equation:

$$\frac{\partial v_z}{\partial z} + \frac{1}{r}\frac{\partial}{\partial r}(rv_r) = 0 \qquad (2.137)$$

EXPERIMENTAL TECHNIQUES

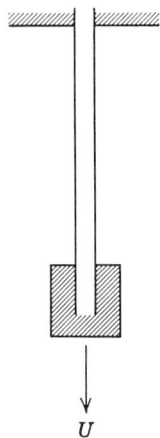

Fig. 2.18. Definition sketch for a simple elongational flow.

In order to satisfy continuity, then, it must be true that

$$\frac{\partial v_z}{\partial z} = \text{constant} = -\frac{1}{r}\frac{\partial}{\partial r}(rv_r) \qquad (2.138)$$

A solution which satisfies the boundary conditions $v_z = 0$ at $z = 0$, $v_z = U$ at $z = L(t)$, and $v_r = 0$ at $r = 0$, is

$$v_z = Uz/L \qquad v_r = -Ur/2L \qquad (2.139)$$

The length of the sample increases from its initial value L_0 according to

$$L(t) = L_0 + \int_0^t U(t)\,dt \qquad (2.140)$$

Since the volume of the sample must remain constant, it follows that the area of the cross section decreases with time, and is given by

$$A(t)/A_0 = L_0/L(t) = 1/\alpha_1 \qquad (2.141)$$

α_1 is called the "principal extension ratio."

The strain rate $\dot{\epsilon}$ is easily seen to be expressible by any of the terms

$$\dot{\epsilon} = \frac{\partial v_z}{\partial z} = \frac{U}{L} = \left(\frac{1}{L}\right)\frac{dL}{dt} = \left(\frac{1}{\alpha_1}\right)\frac{d\alpha_1}{dt} \qquad (2.142)$$

In the usual elongation experiment the velocity U is held constant, say $U = U_0$. Then $\dot{\epsilon}$ becomes

$$\dot{\epsilon} = U_0/L = U_0/\alpha_1 L_0 \qquad (2.143)$$

and it is clear that the strain rate continually decreases as elongation proceeds. Since material functions may depend on $\dot\epsilon$, one must be careful in interpreting results from such an experiment. Smith (38) describes such an elongation experiment on polyisobutylene, and claims to have performed elongation at constant strain rate for α_1 up to 1.5. Since Smith performed experiments at different values of U_0/L_0 over a range of three orders of magnitude, he evidently feels warranted in neglecting a 50% variation in $\dot\epsilon$. Examination of his results would seem to justify this approximation. One should, however, understand that elongation at constant speed *is not* a constant rate of strain experiment, except in approximation at low extension.

A quantitative example of behavior under constant speed extension can be obtained by examining a simple linear viscoelastic model of the form

$$\tau + \lambda \frac{d\tau}{dt} = \eta_0 \Delta$$

In the limit of vanishing relaxation time λ, one obtains the newtonian fluid. For the axial direction, Δ_{11} is $2\dot\epsilon$, and hence is given by $\Delta_{11} = 2U_0/\alpha_1 L_0$. Since $L = L_0 + U_0 t$, it follows that $\alpha_1 = 1 + U_0 t/L_0$. If the time variable is replaced by α_1 in the constitutive equation above and dimensionless variables are introduced, the resulting differential equation may be written as

$$\tau^* + \Lambda \frac{d\tau^*}{d\alpha_1} = \frac{1}{\alpha_1} \tag{2.144}$$

where

$$\tau^* = \tau_{11} L_0 / 2\eta_0 U_0 \quad \text{and} \quad \Lambda = \lambda U_0 / L_0$$

The initial condition is simply $\tau^* = 0$ at $\alpha_1 = 1$. The solution for this model may be represented as τ^* vs. α_1, with Λ as a parameter. Figure 2.19 shows a family of solutions. In the limit of vanishing Λ, the newtonian solution, $\tau^* = 1/\alpha_1$, is approached.

The axial force required to produce constant speed elongation is given by

$$F = T_{11} A = \tfrac{3}{2} \tau_{11}(A_0/\alpha_1) \tag{2.145}$$

This may be rewritten in the form

$$\frac{1}{2A_0\eta_0} \frac{F\alpha_1}{U_0/L_0} = \tfrac{3}{2}\tau^*$$

Hence, data for a given material (and fixed A_0) could be plotted as $F\alpha_1/(U_0/L_0)$ as a function of α_1, and the resulting curves would separate according to the nominal elongation rate U_0/L_0. For long times, however, the newtonian region would be approached, and curves at various elonga-

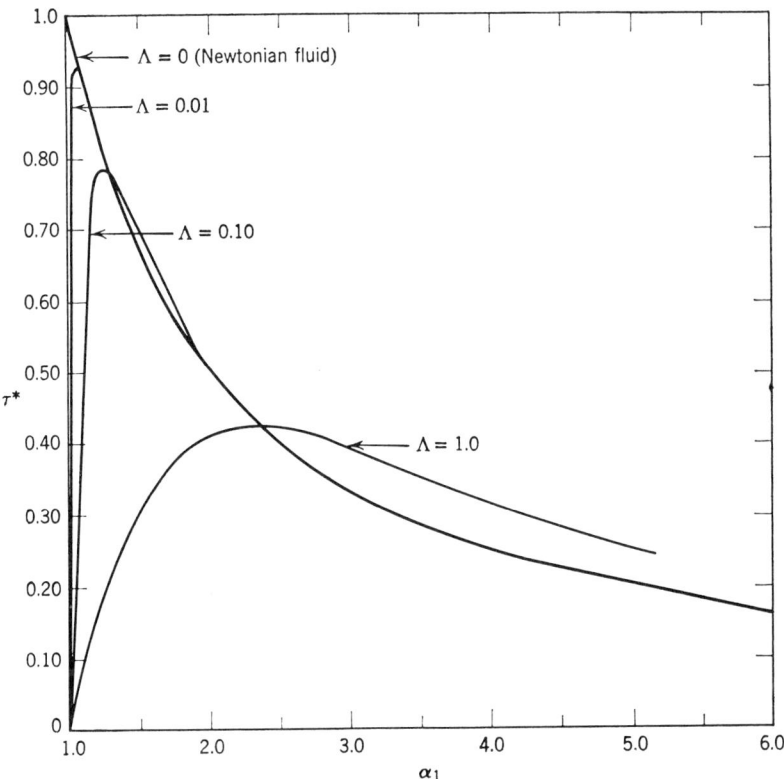

Fig. 2.19. Comparison of newtonian and viscoelastic elongational flows, at constant speed, plotted as reduced stress *vs.* principal extension ratio.

tion rates would converge to a single curve. Figure 2.20 illustrates this behavior.

A constant rate of strain experiment may be achieved by an elongation program given by

$$U = \beta L_0 e^{\beta t} \tag{2.146}$$

where β is some constant. For this choice of U it is found that L is

$$L = L_0 e^{\beta t} \tag{2.147}$$

and

$$\dot{\epsilon} = U/L = \beta \tag{2.148}$$

Hence, by suitable choice of β, the desired value of strain rate may be

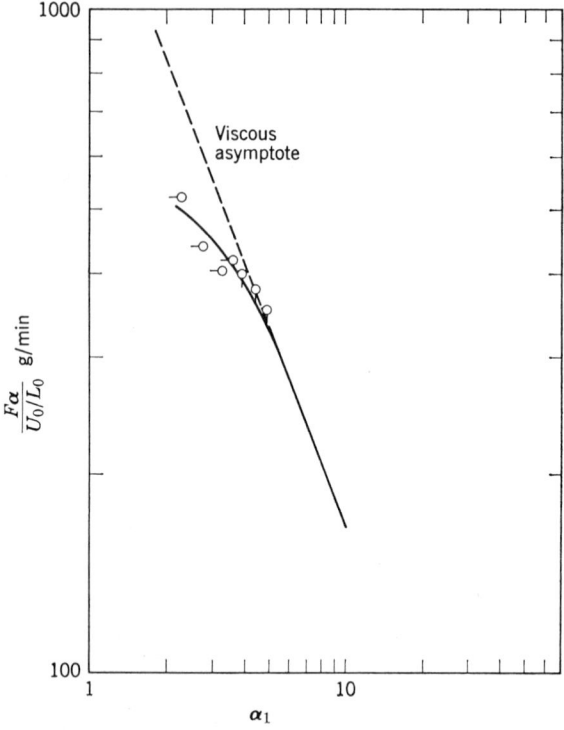

Fig. 2.20. Constant speed elongation data for a viscoelastic fluid. At these slow speeds the final portion of the curve defines a newtonian viscosity. Data shown are for undiluted polyisobutylene (B50) at 25°C. The viscosity is calculated to be 1.2×10^6 poise.

achieved. Ballman (39) presents experimental results obtained with polystyrene rods subjected to elongation at constant rate of strain. After an initial period of stress buildup (elastic response) the stress–elongation curve flattens, as illustrated in Fig. 2.21. Since T_{11} is constant over this portion of the curve and $\dot{\epsilon}$ is constant for the entire experiment, it is possible to calculate the elongational viscosity (see Eq. 1.14) from $\eta_e = T_{11}/\dot{\epsilon}$. In terms of the usually measured quantities this would be

$$\eta_e = F\alpha_1/A_0\dot{\epsilon} \qquad (2.149)$$

Ballman presents η_e as a function of T_{11}, and compares the results with the corresponding data for the shear viscosity η as a function of τ_{12}. Figure 2.22 shows the results. At low deformation stresses the newtonian result $\eta_e = 3\eta_0$ is confirmed. In the region of non-newtonian shear behavior the

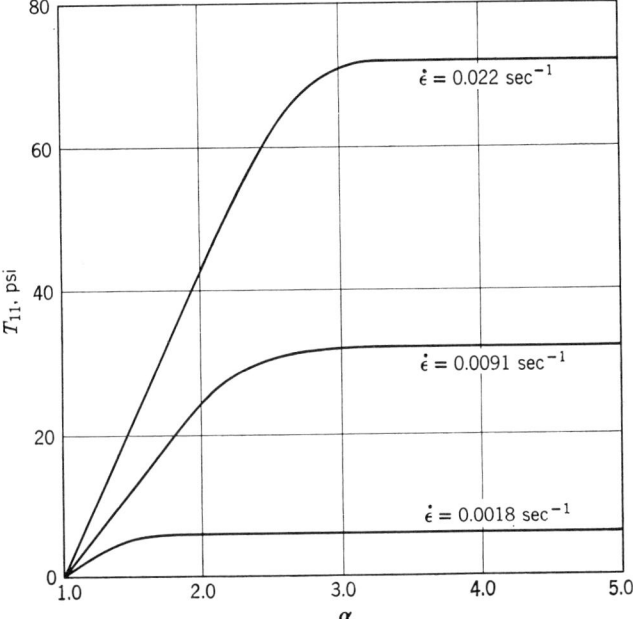

Fig. 2.21. Elongation at constant rate of strain. Data of Ballman (39) for molten polystyrene.

elongational viscosity remains nearly constant while the shear viscosity falls nearly two orders of magnitude. In Chapter 3 it will be seen that viscoelastic theories actually predict an *increase* in η_e with increasing stress (Eq. 3.86).

II. DYNAMIC MEASUREMENTS

The flows discussed up to this point have been those maintained under steady state (time-independent) conditions. It is possible to carry out tests under *dynamic* conditions in which an external variable, such as the pressure drop across a capillary or the angular velocity of one cylinder of a Couette instrument, undergoes some programmed time variation. One then measures some features of the transient response of the fluid and infers therefrom the properties of the fluid under examination.

Two flows which particularly lend themselves to dynamic measurements are the oscillatory flow and the relaxational flow. In the former experiment the system has imposed upon it sinusoidal external conditions of known frequency and amplitude. The frequency, amplitude, and phase of the

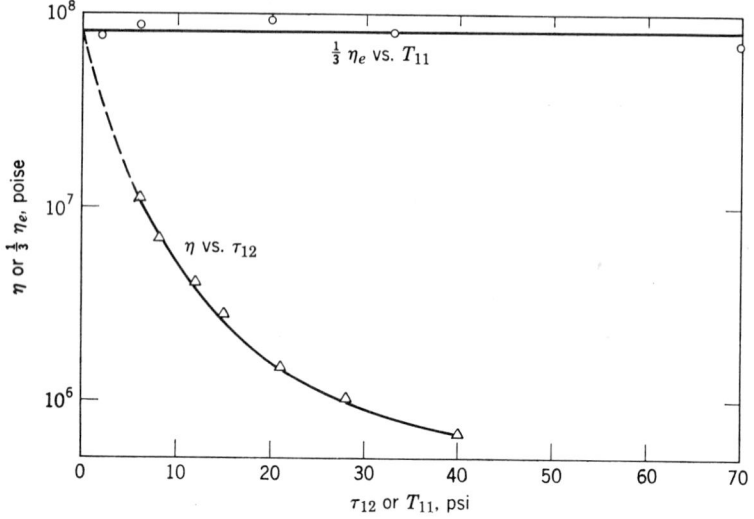

Fig. 2.22. Comparison of shear viscosity with elongational viscosity, based on data of Ballman (39) for molten polystyrene.

response of the fluid are measured. In a relaxational flow external conditions, such as force or strain, undergo a rapid change from one steady state to a second steady state. The response of the fluid as it approaches a new equilibrium state is then measured.

In both types of flows, by appropriate measurements, it is possible in principle to determine the viscosity and the normal stress coefficients for the fluid. If the experiments are properly designed and interpreted these coefficients should be identical with those measured in some steady state configuration. It has been more common, however, to define *new* coefficients which are peculiar to the particular type of dynamic test undertaken. It is important, then, to investigate the possibility that these new coefficients might be related to the viscosity and normal stress coefficients, or otherwise to establish these coefficients as independent properties of the material.

Dynamic experiments are of interest for a number of reasons. Historically, since polymeric fluids are recognized to exhibit time effects such as stress relaxation, it has been natural to examine such effects directly through stress relaxation experiments. Furthermore, a very severe test of the applicability of a proposed constitutive equation is its ability to predict *both* steady state and dynamic behavior. Such tests are often useful in differentiating among models which make similar predictions for steady state flows, but differ significantly in predicted dynamic response.

Finally, it is often possible to achieve a time scale in a dynamic experiment which is not easily accessible to a steady state experiment. Hence dynamic experiments are useful for extending the range of time scale over which a particular material is examined.

A. Oscillatory Flow (40)

Consider a fluid filling the region between long concentric cylinders, as in Fig. 2.3. Suppose the outer cylinder oscillates with frequency $\bar{\omega}$ (sec.$^{-1}$) and amplitude ϑ_0. The angular velocity ω, the shear rate $\dot{\gamma}$, and the shear stress τ will exhibit sinusoidal behavior of frequency $\bar{\omega}$. In general these quantities will not be in phase with the input oscillation. If the inner cylinder is free to respond, for example, its angular velocity might be

$$\omega_i = \Omega_i \cos(\bar{\omega}t + \delta) \tag{2.150}$$

while the input oscillation would be

$$\omega_0 = \Omega_0 \cos \bar{\omega}t \tag{2.151}$$

A convenient representation of such oscillating quantities is in terms of complex exponentials. Equation 2.150 may be written

$$\omega_i = \text{Re}\{\mathbf{\Omega}_i e^{i\bar{\omega}t}\} \tag{2.152}$$

where $\mathbf{\Omega}_i = \Omega_{iR} + i\Omega_{iI}$ is a complex amplitude and Re{ } means "the real part of { }." In terms of these quantities it may be shown that (41)

$$\Omega_i = \sqrt{\Omega_{iR}^2 + \Omega_{iI}^2} \tag{2.153}$$

and

$$\delta = \tan^{-1}(\Omega_{iI}/\Omega_{iR}) \tag{2.154}$$

In the same way one may write the shear stress and shear rate as

$$\tau = \text{Re}\{\mathbf{T}e^{i\bar{\omega}t}\} \tag{2.155}$$

and

$$\dot{\gamma} = \text{Re}\{\dot{\mathbf{\Gamma}}e^{i\bar{\omega}t}\} \tag{2.156}$$

By analogy to earlier definitions one might define a viscosity coefficient as $\eta = \tau/\dot{\gamma}$. Because τ and $\dot{\gamma}$ are not necessarily in phase with one another such a definition leads to a time-dependent viscosity which, then, does not really represent a material property. Instead, it is common to define a complex viscosity coefficient as (42)

$$\eta^* = \mathbf{T}/\dot{\mathbf{\Gamma}} = \eta' - i\eta'' \tag{2.157}$$

The parameter η' is the "dynamic viscosity" and is a function of frequency $\bar{\omega}$ in essentially the same way that the steady shear viscosity η is a function of shear rate. The parameter η'' is found to be a measure of elastic response of the material, and can be shown to be related to the steady shear normal stress coefficient Ψ_{12}. The relationships among η, Ψ_{12}, η', and η'' will be discussed more thoroughly in Chapter 5. It is appropriate at this point to discuss some experimental techniques for the measurement of these dynamic parameters.

A common method of operation (43) has the bob suspended from a thin torsion wire. The outer cylinder, or cup, oscillates and, after initial transients damp out, a sinusoidal motion is induced in the annular body of fluid concentric with the cup and bob. A sinusoidal torque is exerted upon the bob which is opposed by the elasticity of the suspending wire. At equilibrium the bob oscillates with an amplitude and phase which depend upon the fluid properties, the amplitude and frequency of the cup motion, the moment of inertia of the cup, and the torsion constant of the wire.

The equation of motion of the bob may be written as

$$I\ddot{\vartheta}_i = \mathcal{T} - K\vartheta_i \tag{2.158}$$

where I and K are the moment of inertia and the torsion constant mentioned above and $\mathcal{T}$ is the torque exerted on the bob by the fluid.

The equation of motion of the fluid is

$$\rho \frac{\partial r\omega}{\partial t} = \frac{1}{r^2} \frac{\partial r^2 \tau}{\partial r} \tag{2.159}$$

Here v_θ has been replaced by $r\omega$, and τ stands for $\tau_{r\theta}$. If η^* is introduced, the resulting dynamic equation for the fluid is

$$\rho r \frac{\partial \omega}{\partial t} = \frac{1}{r^2} \frac{\partial}{\partial r} r^3 \eta^* \frac{\partial \omega}{\partial r} \tag{2.160}$$

If the angular velocity ω is now replaced by $\partial \vartheta / \partial t$, where ϑ is the angular deflection from equilibrium, and if ϑ is written as

$$\vartheta = \theta e^{i\tilde{\omega} t} \tag{2.161}$$

then Eq. 2.160 becomes, after some algebraic rearrangement

$$\frac{d^2\theta}{dr^2} + \frac{3}{r} \frac{d\theta}{dr} + \beta^2 \theta = 0 \tag{2.162}$$

where

$$\beta^2 = \rho\tilde{\omega}/i\eta^*$$

This result is subject to the assumption that η^* may be considered constant over the region between the cup and bob. This is not a bad assumption as long as the clearance between the two surfaces is small compared to the bob radius, a condition often met in practice.

The solution to Eq. 2.162 is

$$\theta(r) = \frac{1}{r}[AJ_1(\beta r) + BY_1(\beta r)] \qquad (2.163)$$

where the constants A and B are to be found from appropriate boundary conditions. J_1 and Y_1 are Bessel functions which are tabulated in many standard references (44).

The torque at the bob surface is found to be

$$\mathcal{T} = 2\pi R^2 L\eta^*\left(r\frac{\partial \dot{\vartheta}}{\partial r}\right)_{r=R} = 2\pi R^2 L i\bar{\omega}\beta\eta^*[AJ_2(\beta R) + BY_2(\beta R)]e^{i\bar{\omega}t} \qquad (2.164)$$

When this is substituted into Eq. 2.158 the solution for ϑ_i may be obtained. The appropriate boundary conditions are then $\theta(R) = \vartheta_i$ and $\theta(R_0) = \vartheta_0$. After performing the appropriate algebra it is possible to solve for the ratio of ϑ_0 to ϑ_i with the result

$$\frac{\vartheta_0}{\vartheta_i} = 1 + \frac{i}{\eta^*}\left[\frac{I\bar{\omega}-K}{4\pi L\bar{\omega}}\frac{R_0^2-R^2}{R_0^2R^2} + \frac{\bar{\omega}\rho}{8}\frac{(R_0^2-R^2)^2}{R_0^2}\right] + O\left(\frac{1}{\eta^{*2}}\right) \qquad (2.165)$$

Without loss of generality the cup motion may arbitrarily be taken to be strictly cosinusoidal, so that ϑ_0 is strictly a real quantity. The bob response is, in general, out of phase with the cup and so ϑ_i is a complex quantity: $\vartheta_i = \vartheta_R + i\vartheta_I$. The primary measurements consist of the amplitude ratio $q = |\vartheta_i|/\vartheta_0$ and the phase difference $\delta = \tan^{-1}(\vartheta_I/\vartheta_R)$. From these two quantities, at any particular frequency $\bar{\omega}$, the coefficients η' and η'' may be determined. By performing experiments over a range of frequencies, the frequency dependence of η' and η'' may be determined. The analysis outlined above leads to the following results:

$$\eta' = -\frac{Gq \sin \delta}{1 + q^2 - 2q\cos\delta} \qquad (2.166)$$

$$\eta'' = \frac{G(q\cos\delta - 1)}{1 + q^2 - 2q\cos\delta} \qquad (2.167)$$

where

$$G = \frac{I\bar{\omega}^2 - K}{4\pi L\bar{\omega}}\frac{R_0^2 - R^2}{R_0^2 R^2} + \frac{\bar{\omega}\rho}{8}\frac{(R_0^2-R^2)^2}{R_0^2}$$

It is also possible to perform an experiment in which the cup is fixed and the upper end of the torsion wire is twisted periodically through an angle $\psi = \psi_0 \cos \bar{\omega}t$. The bob then performs periodic oscillations of frequency $\bar{\omega}$ but of different amplitude and phase, so that $\vartheta_i/\psi_0 = qe^{-i\delta}$. Again, measurement of the amplitude ratio q and the phase angle δ

permits calculation of η' and η''. For this case the results are (43)

$$\eta' = \frac{R_0^2 - R^2}{4\pi L R_0^2 R^2} \left[I\tilde{\omega}^2 - K + \frac{K \cos \delta}{q} \right] \tag{2.168}$$

$$\eta'' = \frac{R_0^2 - R^2}{4\pi L R_0^2 R^2} \frac{K \sin \delta}{q} \tag{2.169}$$

subject to the (usually valid) approximation $\rho\tilde{\omega}R_0^2/|\eta^*| \ll 1$.

An example of the measurement of dynamic properties may be found in a study of molten polystyrene by Cox, Nielsen, and Keeney (45). The sample was contained in the annular space between two cylinders of radii 6 and 8 mm. and height 4 cm. The cup oscillated at frequencies in the range 0.05–5 cps, with a maximum amplitude of about $\frac{1}{2}°$. The amplitude ratio and phase difference were measured electronically over the frequency range, and η' and η'' were calculated from Eqs. 2.166 and 2.167. Figure 2.23 shows typical behavior for $\eta'(\tilde{\omega})$. Instead of η'', a related function $G' = \tilde{\omega}\eta''$, the "dynamic rigidity," is shown as a function of $\tilde{\omega}$. If η'' were plotted it would be seen to go through a maximum.

It is common practice among those who investigate "solidlike" materials to define material coefficients based upon the shear *strain* γ rather than the shear *rate* $\dot{\gamma}$. In a sinusoidal experiment γ would be written as

$$\gamma = \text{Re}\{\Gamma e^{i\tilde{\omega}t}\} \tag{2.170}$$

But, since $\dot{\gamma} = d\gamma/dt$, γ could also be written in the form

$$\gamma = \text{Re}\{(\dot{\Gamma}/i\tilde{\omega})e^{i\tilde{\omega}t}\} \tag{2.171}$$

A "complex shear modulus" is defined as $\mathbf{G}^* = \mathbf{T}/\mathbf{\Gamma}$. Since $\dot{\Gamma} = i\tilde{\omega}\Gamma$ it follows that $\mathbf{G}^* = i\tilde{\omega}\boldsymbol{\eta}^*$. $\mathbf{G}^*$ is usually written as $G' + iG''$, where G' is called the "storage modulus," or "dynamic rigidity," and G'' is called the "loss modulus." As noted above, $G' = \tilde{\omega}\eta''$. It can be seen also that $G'' = \tilde{\omega}\eta'$.

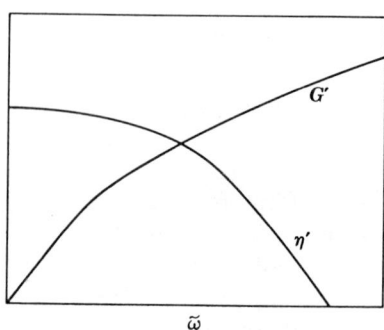

Fig. 2.23. Typical shapes for the dynamic functions η' and G' as a function of frequency $\tilde{\omega}$.

A "complex compliance" J^* may also be defined as $J^* = J' - iJ'' = \Gamma/T$. By simple manipulation it can be shown that $J' = G'/(G'^2 + G''^2)$ and $J'' = G''/(G'^2 + G''^2)$. J' and J'' are called the storage and loss compliances, respectively.

Many other methods are available for the measurement of the dynamic coefficients defined here. A detailed review of experimental aspects is contained in the article by Ferry (42). A typical recent example of an experimental investigation of dynamic properties is the paper of Philippoff (46).

It is possible to define dynamic normal stress coefficients by analogy to the definition of the complex viscosity, and to analyze experimental systems suitable to the measurement of such coefficients (47). As yet, no such data have appeared in the literature.

B. Creep and Relaxation

In the Introduction a comment was made to the effect that a given material might show different classes of mechanical response under conditions involving different time scales. Consider, for example, a sheet of polymer suddenly subjected to a constant shear stress τ by the action of a force pulling on one plate bonded to the polymer, as in Fig. 2.24. Two limiting responses are possible. If the material is a true elastic solid, then the shear strain γ immediately attains and remains at a constant value as long as the stress is maintained. If the stress is removed the strain relaxes instantaneously and the solid returns to its initial unstressed state. If the material, however, is a fluid, then the shear strain continuously increases with time as long as the stress is maintained. If the stress is suddenly removed the strain does not relax, but instead remains constant. An irreversible deformation, flow, has occurred. Figure 2.25 shows these responses.

In most polymeric materials neither of these simple responses occurs. Instead, a combination of elastic and viscous behavior gives rise to the strain history indicated in Fig. 2.26. One associates such behavior with viscoelastic materials.

If the material is observed over a time scale very short compared with

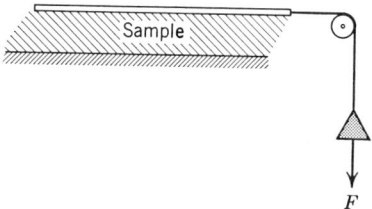

Fig. 2.24. Definition sketch for a shear creep experiment at constant shear stress.

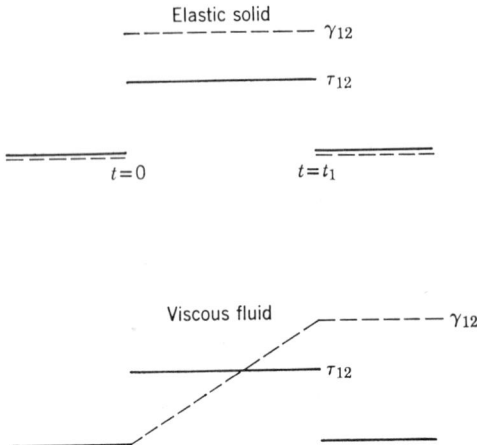

Fig. 2.25. Hypothetical shear strain response to a constant shear stress imposed for a finite time t_1. The perfectly elastic solid shows instantaneous and complete recovery of strain upon release of stress. The purely viscous fluid shows no recovery of strain.

the time λ of Fig. 2.26 it is likely that the observer would conclude that the material is an elastic solid. On the other hand, if the time λ is very short compared with the response time of suitable measuring instruments, one is likely to conclude that the material under examination is a fluid. For many polymeric materials, at least in certain ranges of temperature and molecular weight, the complete response curve is accessible to measurement, and so viscoelastic parameters may be measured.

The response curve of Fig. 2.26 may be described by a function called the "creep compliance," defined as

$$J(t) = \gamma(t)/\tau \tag{2.172}$$

It is common (48) to separate the creep compliance into three parts:

$$J(t) = J_0 + J\psi(t) + t/\eta \tag{2.173}$$

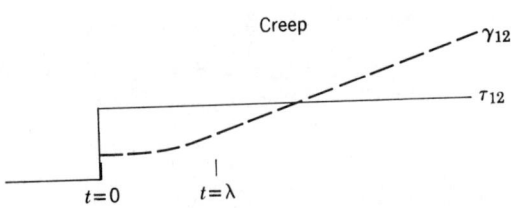

Fig. 2.26. Hypothetical shear strain response to a step change in shear stress for a viscoelastic fluid.

J_0 is the "ordinary elastic compliance" and represents the initial *elastic* response to stress. The term t/η represents the long time *flow* response. $\psi(t)$ is called the "retarded elasticity function," and is defined in such a way that $\psi(t)$ approaches unity as time increases indefinitely. J is the "steady state shear compliance," and like J_0 and η is a material constant, for a given shear stress τ. The retarded elasticity function, while not a material *constant*, is nevertheless a characteristic property of a material. An experiment giving rise to a response such as is shown in Fig. 2.26 is usually referred to as a "creep experiment."

If the material is in elastic equilibrium by virtue of some previous creep behavior followed by steady flow, and if the stress causing this flow is suddenly removed, then retarded elastic recovery occurs, as shown in Fig. 2.27. The strain relaxation history is then the reverse of the creep curve.*
If $\gamma(0)$ is the strain at the instant of stress removal, then

$$\gamma(0) - \gamma(t) = \tau[J_0 + J\psi(t)] = \tau[J(t) - t/\eta] \qquad (2.174)$$

where τ represents the constant stress which led to $\gamma(0)$. Note that in strain recovery there is no flow mechanism since the external stress no longer acts on the material. An obvious test of consistency of any *creep*

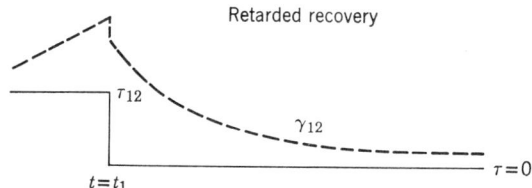

Fig. 2.27. Hypothetical strain recovery upon release of a shear stress for a viscoelastic fluid.

experiment is the prediction of the *recovery* curve from measurements, under creep, of J_0, J, and $\psi(t)$.

The work of van Holde and Williams (49) provides one example of a good creep experiment. The measurements were made with a "parallel plate viscoelastometer," shown schematically in Fig. 2.28. The primary measurement was the horizontal displacement of the thin plate sandwiched by the samples. The polymer studied was undiluted polyisobutylene, a rubbery solid in the range of molecular weights used by these authors. The samples were disks $\frac{1}{4}$ in. thick by 1 or $\frac{3}{4}$ in. in diameter, and were "tacky" enough to be held in place by slight compression of the frame of the apparatus.

---

* This assumes that irreversible breakdown of chemical structure, or development of crystallinity, do not result from the strain imposed on the material.

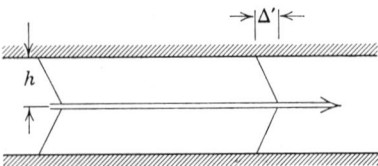

Fig. 2.28. A "parallel plate viscoelastometer" for the measurement of shear creep (49).

Figure 2.29 shows both creep and recovery data from one experiment. The deflection was measured photographically, and the ordinate shown must be converted to the actual strain using $\gamma = \Delta' f/h$ where f is the reciprocal of the magnification of the photographic image and h is the thickness of the sample.

The dashed line of Fig. 2.29 is the recovery curve that is predicted from the creep curve by considering $J(t) - t/\eta$ (hence the label "creep-flow"). The recovery curve is plotted as $\Delta'(0) - \Delta'$, the deflection at the instant of stress removal minus the instantaneous deflection at any time thereafter. The agreement is seen to be good.

From the long time portion of the creep curve the viscosity of the material may be obtained. Since it is practically impossible to measure viscosity of such rubberlike materials by simple techniques such as capillary flow or cone and plate shearing, which must be restricted to relatively low viscosities, the creep experiment provides an important extension of viscometry methods to the region of extreme viscosities. van Holde and Williams measured viscosities in the range 10^6–10^9 poise. Because the time scale of the creep experiment is so low these viscosities are "zero-shear" viscosities; the polymer is essentially newtonian in its flow behavior under these conditions.

If higher shear stresses are applied it is possible to measure the non-newtonian viscosity of a "solid" polymer. van Holde and Williams worked with shear stresses less than 10^4 dynes/cm.^{2}. A more recent study by Gibbs and Merrill (50), using a similar parallel plate arrangement but accommodating shear stresses up to 10^6 dynes/cm.^{2}, shows clear evidence of non-newtonian flow.

Retarded elastic behavior may also be studied in a coaxial cylinder arrangement. The bob has a constant torque applied to it and the angular deflection of the bob is measured as a function of time. From the long time response the viscosity of the polymer may be calculated by replacing Ω in Eq. 2.47 by $d\theta/dt$, where θ is the angular deflection of the bob, and by using Eq. 2.26 with τ_R replaced by $\eta\dot{\gamma}$. Under most conditions of operation the newtonian approximation implied here would not be in error significantly.

Upon removal of torque the elastic recovery may be followed and

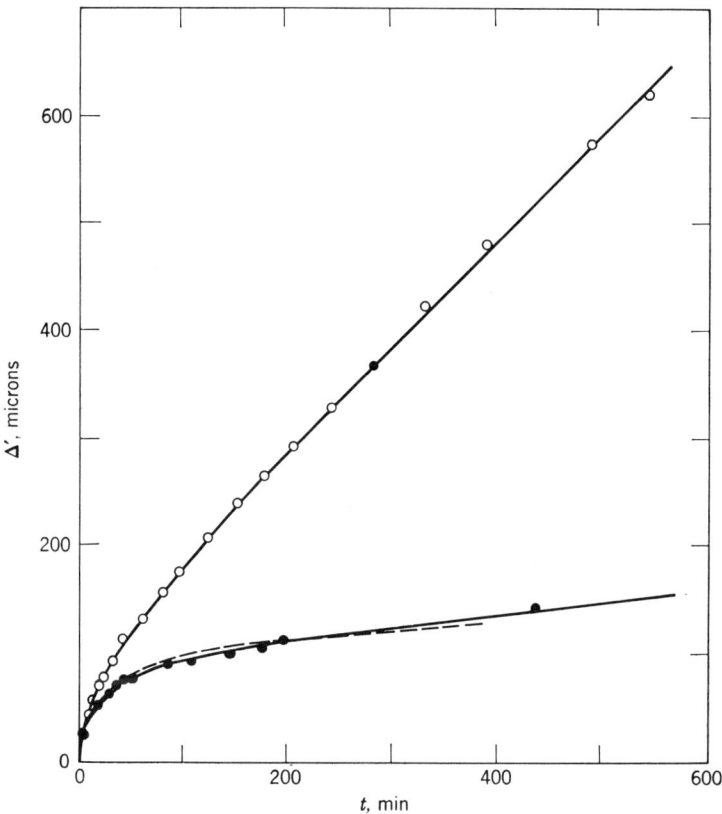

Fig. 2.29. Shear creep and recovery of polyisobutylene, a soft rubbery elastomer.

viscoelastic parameters calculated from (48)

$$\theta(0) - \theta(t) = \frac{4\pi R^2 L \mathcal{T}}{1 - s^2} [J_0 + J\psi(t)] \qquad (2.175)$$

Eq. 2.175 is valid only after elastic equilibrium (the steady flow portion of the creep curve) has been achieved.

Leaderman, Smith, and Jones (51) present a detailed discussion of the use of such an instrument for the measurement of properties of polyisobutylene. The reader interested in some of the experimental details of such an investigation should consult this reference.

Another instrument used to measure viscoelastic properties is the torsional pendulum. A circular disk hung from a torsion wire is in contact with the material to be studied. The disk is displaced through a small angle about its axis, the torsion in the wire tends to restore the disk to its equilibrium position, but the inertia of the disk causes it to overshoot the equilibrium position. Thus the disk performs torsional oscillations of fixed frequency and damped amplitude. The frequency and damping factor of the oscillations are measured, and may be related to instrument constants and properties of the material.

Plazek, Vrancken, and Berge (52) present data obtained with such a torsional instrument. In addition to the oscillatory experiment, their instrument can be used in a creep experiment. In this case a constant torque acts on the disk (without a torsion wire) and the resulting angular deflection is monitored as a function of time. In a later modification (53) of this instrument Plazek measures viscosities of the order of 10^{11} poise for polyisobutylene, and presents creep and recovery curves for the same material.

The preceding experiments, and the corresponding dynamic coefficients, are related to shear behavior. Elastic materials are commonly studied in tension as well as in shear, and it is common to study viscoelastic materials in tensile experiments. Thus one may study elongation (creep) under constant tensile load or tensile stress relaxation at constant tensile strain. Corresponding to the former experiment one may define a tensile creep modulus as $E_c(t) = T_{11}/\epsilon(t)$, where $\epsilon(t)$ is the strain (change in length/initial length) resulting from the constant tension T_{11}. For the latter experiment a relaxation modulus $E_r(t) = T_{11}(t)/\epsilon$ may be defined.

A simple relaxation experiment involves the sudden stretching of a strip of material to a fixed elongation. The tendency for creep is counteracted by continually reducing the force maintaining the elongation in such a way that the initial elongation is not altered. From the resulting data of stress vs. time it is possible to calculate $E_r(t)$. An example of such data is shown in Fig. 2.30, taken from the study of Catsiff and Tobolsky (54).

As in the case of shear creep, tensile creep experiments may be used to measure the viscosity of semisolid materials. Karam and Bellinger (55) present such data for polystyrene at temperatures in the range 60–240°C. Viscosities as high as 10^{15} poise were measured. Tensile creep is similar to the simple elongation experiment described on pp. 60–65. The difference lies in the fact that in elongational flows the strain is programmed and the measured response is the stress (force), while in the creep experiment the force is usually fixed and the measured response is the strain.

C. Wave Propagation on a Jet

Stress relaxation experiments are generally performed with elastomers having relaxation times* of the order of a tenth of a second or more.

---

* Relaxation time is used in the qualitative sense here. A suitable definition might be the time required for the stress to fall to 5% of its initial value, upon release of a previously maintained constant strain.

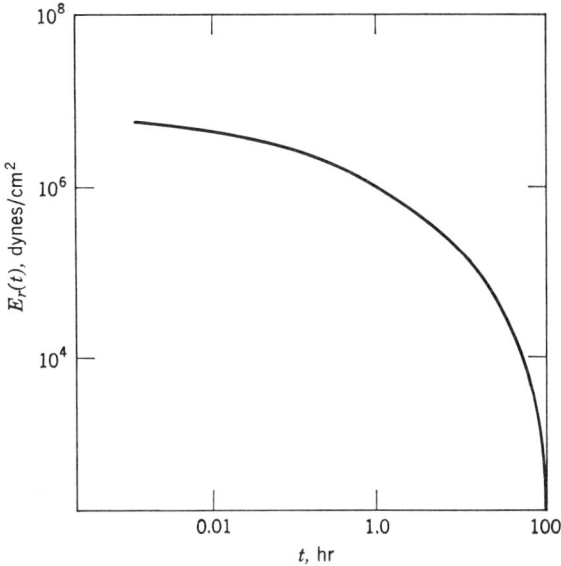

Fig. 2.30. The tensile relaxation modulus of polyisobutylene.

Relaxation within a shorter time is difficult to measure in standard relaxation tests. Hence, stress relaxation in liquids, which generally have small relaxation times, must be examined by special techniques. One such technique involves wave propagation on a jet of liquid issuing from a long capillary.

The principle of this method is based upon the observation (56) that a liquid issuing from a capillary which vibrates sinusoidally perpendicular to its axis will have superimposed on its trajectory a transverse wave. Figure 2.31 shows a photograph of such a wave pattern. For a newtonian liquid, analysis of the wave pattern yields information relating the surface tension of the liquid to the wavelength and nodal length of the wave, the flight velocity of the jet, and the frequency of vibration of the tube (57).

Analysis of the corresponding viscoelastic problem is more complex because of stress relaxation in the jet. The fluid, while passing through the capillary, develops an axial normal stress. If the tube is sufficiently long this stress will be in equilibrium with the shear field in the tube. This is the same stress discussed in connection with jet expansion and thrust experiments [see pp. 42–47]. When the fluid issues from the tube the wall shear stress vanishes and the normal stress is free to relax. If, now, a wave is propagating along the jet, it travels through a region of rapidly relaxing tension. The varying tension leads to a varying nodal length, from which the rate of stress relaxation may be determined.

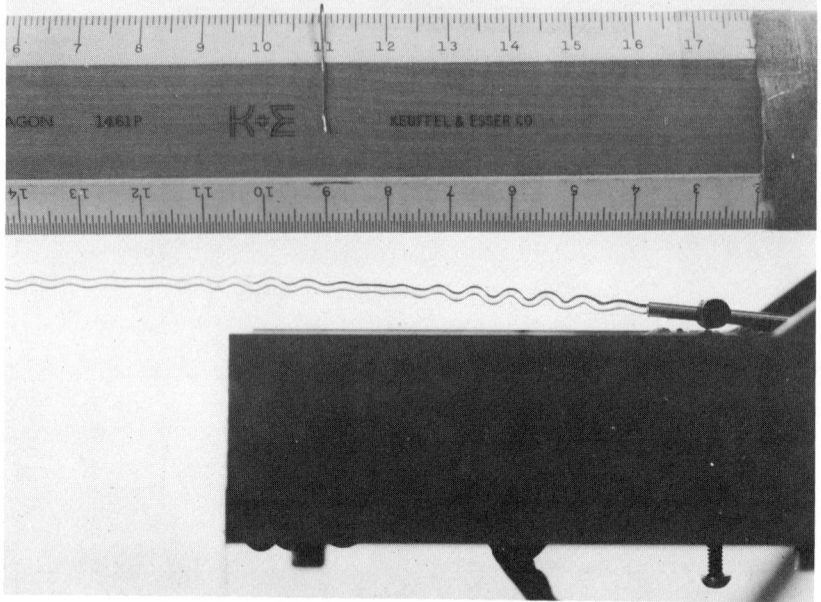

Fig. 2.31. A transverse wave imposed on a jet issuing from a capillary.

The theoretical analysis takes as a starting point a study of wave propagation presented by Goren and Gavis (58). The wave equation for the jet is given as

$$\frac{\partial^2 \psi}{\partial t^2} + 2V \frac{\partial^2 \psi}{\partial z\, \partial t} + (V^2 - \langle T_{11} \rangle'/\rho) \frac{\partial^2 \psi}{\partial z^2} = 0 \qquad (2.176)$$

subject to the boundary conditions

$$\psi(0, t) = \psi_0 \cos 2\pi \nu t$$

and

$$\frac{D\psi(0, t)}{Dt} = \frac{\partial \psi}{\partial t} + V \frac{\partial \psi}{\partial z} = -\psi_0 2\pi \nu \sin 2\pi \nu t$$

where ψ is the wave amplitude at position z ($z = 0$ is the tube exit) and time t. $\langle T_{11} \rangle'$ is the total axial stress *averaged* over the jet cross section. The amplitude of the nodal envelope was shown to obey an ordinary differential equation of the form

$$\frac{d^2 M}{dz^2} + \left(\frac{2\pi \nu}{V}\right)^2 \frac{\langle T_{11} \rangle'}{\rho V^2} M = 0 \qquad (2.177)$$

$$M(0) = 0 \qquad M'(0) = \psi_0 (2\pi \nu)/V$$

The nodal envelope $M(z)$ is defined in Fig. 2.32.

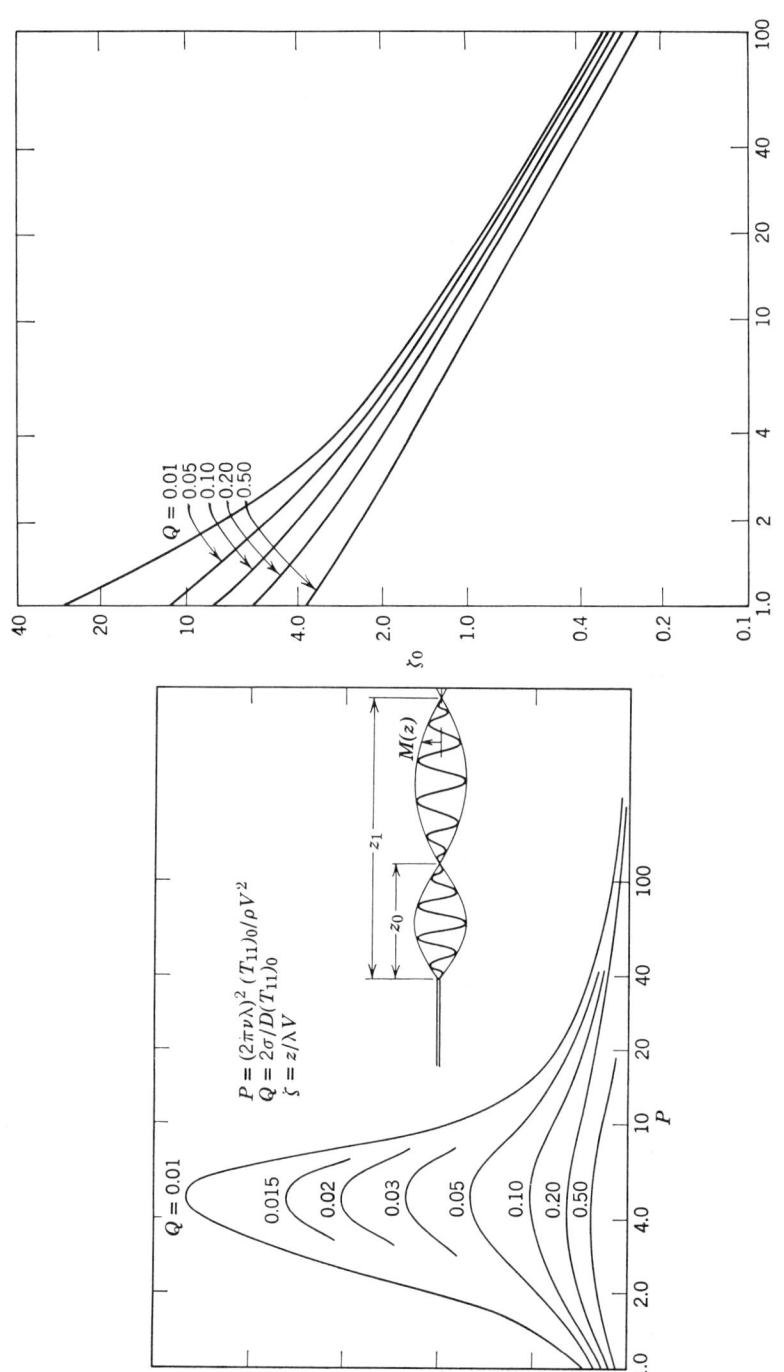

Fig. 2.32. Theoretical solutions for the nodes of a transverse wave propagating on a jet with exponentially relaxing tension.

If $\langle T_{11}\rangle'$ is known as a function of z then $M(z)$ may be determined. The zeros of $M(z)$ are the observed nodal points of the wave pattern, and are measurable from a photograph. In most experimental situations the nodal lengths beyond the second node from the exit become nearly constant. Hence, any accurate information about stress relaxation must be extracted from the first two nodal lengths. (The wavelengths, to good approximation, are independent of material properties and contribute no information.)

One possible procedure (59) for determining the relaxation behavior starts with an assumed model for stress relaxation. For example, exponential decay may be assumed:

$$\langle T_{11}\rangle' = \sigma/R_j + \langle T_{11}\rangle_0 \exp(-z/V_j\lambda) \quad (2.178)$$

Here σ/R_j represents the contribution of surface tension to the stress under which the wave is propagated (57). $\langle T_{11}\rangle_0$ is the axial normal stress at the tube exit, averaged over the tube cross section. V_j is the flight velocity of the jet and λ is the relaxation time of the fluid, assuming exponential decay.

Under usual experimental conditions the radius change of the jet occurs over a distance which is small compared to a nodal length. Hence R_j and V_j should be based upon jet values, rather than capillary values. The relationships, of course, are $R_j = \chi R$ and $V_j = \langle V\rangle/\chi^2$, with R, $\langle V\rangle$, and χ the values used in the discussion of jet expansion on pp. 42–45.

If the exponential model is used solutions of the nodal equation for M may be carried out. Figure 2.32 shows the theoretical results, plotted in terms of dimensionless groups defined in the figure. The ratio of the first two nodal distances is seen to go through a maximum. For very large values of P (long relaxation time), the stress relaxes so slowly that the first two nodes are under nearly the same stress, and so are nearly identical in length. At the opposite extreme the same behavior occurs; this time both nodes are under essentially the same tension because the stress relaxes almost entirely in a distance small compared to the first nodal length.

Since P and Q both involve the unknown $\langle T_{11}\rangle_0$ a trial and error determination is usually involved. Experimental data obtained in a preliminary study (59) with a water solution of carboxymethylcellulose are shown in Fig. 2.33. The relaxation time is not constant, an indication that simple exponential decay is an idealization. No independent measure of λ is available for comparison, although the magnitude is of the expected order.

Values of $\langle T_{11}\rangle_0$ are compared with those calculated from a jet expansion experiment performed on the same material. Equation 2.93 was used for the calculation of $\langle T_{11}\rangle_0$. It was not possible to attain an overlap of the data obtained by the two techniques. Jet expansion measurements become

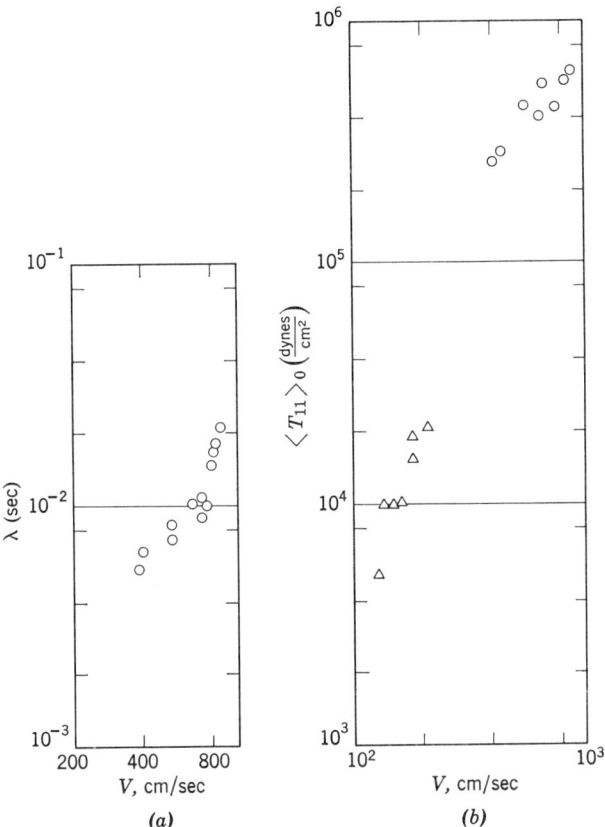

Fig. 2.33. Experimental values for relaxation time and stress, assuming exponential decay, for an aqueous solution of carboxymethylcellulose (CMC).

quite inaccurate at large values of V_j, where χ begins to approach its asymptotic value χ_∞ (Eq. 2.95). But even larger values of flight velocity are required before an accurate wave propagation experiment can be performed. Hence the gap between the two sets of data is practically unavoidable.

It is difficult to argue for or against a single curve through the entire set of data available. The slope of such a curve would be in the neighborhood of three at low shear rates, which is out of line with low shear data on most polymer solutions (24), for which a slope of two is most often observed. It is clear that the wave propagation technique will have to be exploited more carefully, and with more realistic relaxation models, before its utility can be decided.

A significant practical restriction on the wave experiment, pointed out by Goren and Gavis, lies in the demand that $\langle T_{11}\rangle_0/\rho V_j^2$ be less than unity. If this is not the case the character of the wave propagation changes and an interference pattern (i.e., a nodal wave) does not occur. Such behavior has been observed in relatively low-speed jets of highly elastic solutions (21). The nodal equation given above (Eq. 2.177) holds, in fact, only if $\langle T_{11}\rangle_0/\rho V_j^2 \ll 1$. Otherwise distinct nodes do not occur, but standing wave minima appear which are difficult to measure with the required accuracy. The severity of this restriction lies primarily in the fact that operating conditions which give a value of P in the region where ζ_1/ζ_0 goes through a maximum (and so can be more accurately measured) may be in conflict with the restriction that $\langle T_{11}\rangle_0/\rho V_j^2 \ll 1$.

REFERENCES

1. Van Wazer, J. R., J. W. Lyons, K. Y. Kim, and R. E. Colwell, *Viscosity and Flow Measurement*, Interscience, New York, 1963.
2. Bird, R. B., W. E. Stewart, and E. N. Lightfoot, *Transport Phenomena*, Wiley, New York, 1960.
3. Hildebrand, F. B., *Advanced Calculus for Applications*, Prentice-Hall, Englewood Cliffs, N.J., 1962, p. 360.
4. Fredrickson, A. G., *Principles and Applications of Rheology*, Prentice-Hall, Englewood Cliffs, N.J., 1964, pp. 197–199.
5. Coleman, B. D., and W. Noll, *Arch. Ratl. Mech. Anal.*, **3**, 289 (1959).
6. Krieger, I. M., and H. Elrod, *J. Appl. Phys.*, **24**, 134 (1953).
7. Slattery, J. C., *J. Colloid. Sci.*, **16**, 431 (1961).
8. Bird, R. B., *Soc. Plastics Engrs. J.*, **11**, 35 (1955).
9. Toor, H. L., *Trans. Soc. Rheol.*, **1**, 177 (1957).
10. Siegel, R., E. M. Sparrow, and T. M. Hallman, *Appl. Sci. Res.*, **A7**, 386 (1958).
11. Brinkman, H. C., *Appl. Sci. Res.*, **A2**, 120 (1951).
12. Gerrard, J. E., F. E. Steidler, and J. K. Appeldoorn, *Ind. Eng. Chem. Fundamentals*, **5**, 260 (1966).
12a. Kataoka, T., and S. Ueda, *J. Polymer Sci. A*, **3**, 2947 (1965).
13. Turian, R. M., and R. B. Bird, *Chem. Eng. Sci.*, **18**, 689 (1963).
14. Turian, R. M., *Chem. Eng. Sci.*, **20**, 771 (1965).
15. Powell, R. L., and S. Middleman, *Intern. J. Eng. Sci.*, to appear (1968).
16. Truesdell, C. A., and W. Noll, "Non-Linear Field Theories of Mechanics," *Encyclopedia of Physics*, Vol. III/3, S. Flügge, Ed., Springer, Berlin, 1965, p. 458.
17. Williams, M. C., *Chem. Eng. Sci.*, **20**, 693 (1965).
18. Shertzer, C. R., and A. B. Metzner, *Proc. 4th Intern. Congr. Rheol.*, Pt. 2, E. H. Lee, Ed., Interscience, New York, 1965, p. 603.
19. Middleman, S., and J. Gavis, *Phys. Fluids*, **4**, 963 (1961); Gavis, J., and S. Middleman, *J. Appl. Polymer Sci.*, **7**, 493 (1963).
20. Middleman, S., and J. Gavis, *Phys. Fluids*, **4**, 355 (1961).
21. Powell, R. L., Ph.D. Thesis, University of Rochester, Rochester, New York (1967).
22. Gavis, J., and M. Modan, *Phys. Fluids*, **10**, 487 (1967).
23. Huppler, J. D., *Trans. Soc. Rheol.*, **9**:2, 273 (1965).
24. Ginn, R. F., and A. B. Metzner, *Proc. 4th Intern. Congr. Rheol.*, Pt. 2, E. H. Lee, Ed., Interscience, New York, 1965, p. 583.

25. Savins, J. C., *A.I.Ch.E. J.*, **11**, 673 (1965).
26. Metzner, A. B., and G. Astarita, *A.I.Ch. J.*, **13**, 550 (1967). See also Lindgren, E. R., and J.-L. Chao, *Phys. Fluids*, **10**, 667 (1967) and Smith, K. A., et al., *Chem. Eng. Sci.*, **22**, 619 (1967).
27. Hayes, J. W., and R. I. Tanner, *Proc. 4th Intern. Congr. Rheol.*, Pt. 3, E. H. Lee, Ed., Interscience, New York, 1965, p. 389.
28. Miller, C. E., and W. H. Hoppmann, II, *Proc. 4th Intern. Congr. Rheol.*, Pt. 2, E. H. Lee, Ed., Interscience, New York, 1965, p. 619; Hoppmann, W. H. II, and C. E. Miller, *Trans. Soc. Rheol.*, **7**, 181 (1963); Hoppmann, W. H. II, and C. N. Baronet, *ibid.*, **9**:1, 417 (1965).
29. Giesekus, H., *Proc. 4th Intern. Congr. Rheol.*, Pt. 1, E. H. Lee, Ed., Interscience, New York, 1965, p. 249.
30. Williams, M. C., Ph.D. Thesis, University of Wisconsin, Madison, Wisconsin (1964); available from University Microfilms, Inc., Ann Arbor, Michigan.
31. Greensmith, H. W., and R. S. Rivlin, *Trans. Roy. Soc. (London)*, **A245**, 399 (1953).
32. Kotaka, T., M. Kurata, and M. Tamura, *J. Appl. Phys.*, **30**, 1705 (1959).
33. Markovitz, H., *Trans. Soc. Rheol.*, **1**, 37 (1957).
34. Padden, F. J., and T. W. DeWitt, *J. Appl. Phys.*, **25**, 1086 (1954).
35. Markovitz, H., *J. Polymer Sci. B*, **3**, 3 (1965).
36. Markovitz, H., *Proc. 4th Intern. Congr. Rheol.*, Pt. 1, E. H. Lee, Ed., Interscience, New York, 1965, p. 189; Markovitz, H., and D. R. Brown, *Trans. Soc. Rheol.*, **7**, 137 (1963).
37. Coleman, B. D., H. Markovitz, and W. Noll, *Viscometric Flows of Non-Newtonian Fluids*, Springer, New York, 1966, pp. 82–83.
38. Smith, T. L., *J. Polymer Sci.*, **20**, 89 (1956).
39. Ballman, R. L., *Rheol. Acta*, **4**, 137 (1965).
40. Oka, S., in *Rheology*, Vol. 3, Eirich, Ed., Academic Press, New York, 1960, Ch. 2.
41. Wylie, C. R., Jr., *Advanced Engineering Mathematics*, McGraw-Hill, New York, 1960, p. 530.
42. Ferry, J. D., in *Rheology*, Vol. 2, Eirich, Ed., Academic Press, New York, 1960, Ch. 11.
43. Markovitz, H., *J. Appl. Phys.*, **23**, 1070 (1952).
44. Jahnke, E., F. Ende, and F. Losch, *Tables of Higher Functions*, 6th ed., McGraw-Hill, New York, 1960.
45. Cox, W. P., L. E. Nielsen, and R. Keeney, *J. Polymer Sci.*, **26**, 365 (1957).
46. Philippoff, W., *Trans. Soc. Rheol.*, **8**, 117 (1964).
47. Williams, M. C., and R. B. Bird, *Ind. Eng. Chem. Fundamentals*, **3**, 42 (1964).
48. Leaderman, H., *J. Polymer Sci.*, **13**, 371 (1954).
49. Holde, K. van, and J. W. Williams, *J. Polymer. Sci.*, **11**, 2431 (1953).
50. Gibbs, D. A., and E. W. Merrill, *Proc. 4th Intern. Congr. Rheol.*, Pt. 2, E. H. Lee, Ed., Interscience, New York, 1965, p. 183.
51. Leaderman, H., R. G. Smith, and R. W. Jones, *J. Polymer Sci.*, **14**, 47 (1954).
52. Plazek, D. J., M. N. Vrancken, and J. W. Berge, *Trans. Soc. Rheol.*, **2**, 39 (1958).
53. Plazek, D. J., *ibid.*, **7**, 61 (1963).
54. Catsiff, E., and A. V. Tobolsky, *J. Colloid. Sci.*, **10**, 375 (1955).
55. Karam, H. J., and J. C. Bellinger, *Trans. Soc. Rheol.*, **7**, 61 (1964).
56. Gill, S. J., and J. Gavis, *J. Polymer Sci.*, **20**, 287 (1056); **21**, 353 (1956).
57. Middleman, S., and J. Gavis, *Phys. Fluids*, **8**, 222 (1965).
58. Goren, S., and J. Gavis, *ibid.*, **4**, 575 (1961).
59. Middleman, S., and E. Rautenbach, Paper presented to 58th Annual Meeting of A.I.Ch.E., Philadelphia, Dec. 1965.

3

Constitutive Equations—
Continuum Theories

In Chapter 2 it was seen how one may, by suitable experiment, determine material functions, such as viscosity, the normal stress coefficients, or the complex shear compliance. Indeed, one may define a great variety of material functions and measure them by appropriate means. The resultant morass of material functions has led some members of the rheology fraternity to seek some organization in the theory of classification of material response.

That degree of organization which currently exists has come about largely through theoretical studies of continuum mechanics which have led to very broad and general classifications of constitutive equations. As is common among observers of a science which is going through an age of introspection, there is a certain lack of understanding of the value of research into such questions as "What is a simple fluid?" or "How many material functions are required to describe a laminar shear flow of a simple fluid?"

While there is some room for debate on this point it might be said that the major contributions of these fundamental investigations have been the demonstration of an underlying unity among various theories and the delineation of the types of constitutive equations with which one can hope to describe real material response.

In this chapter the results of the continuum approach to the development of constitutive equations will be described. At best, the equations which result yield predictions for the dependence of material functions on shear rate and (sometimes) other kinematic variables. The constants appearing in these equations must be determined through comparison with suitable experimental data. The ultimate test of these theories (indeed, of any theory) lies in their ability to describe accurately the response of some real material to a wide variety of imposed conditions. Ideally one would hope that a given theory would not only predict viscosity, normal stresses, stress relaxation, complex moduli, etc., for some material, but would also do so *accurately*, from a single set of coefficients determined through some one experiment.

All too often, however, one finds that constants chosen to give a good fit to viscosity data do not accurately predict, for example, the normal stress functions. Despite this pessimistic tone let us turn to a consideration of the successes of continuum mechanics in describing the response of real materials.

I. GENERAL CONSIDERATIONS OF CONSTITUTIVE THEORIES

A number of theoretical inquiries have been made into the general question of restrictions which can be placed upon a constitutive equation without regard for the class of behavior one wishes to model. Truesdell reviews these restrictions (1). A particularly important demand is embodied in the so-called "principle of material objectivity." This principle states that a constitutive equation should be of a form independent of the coordinate system selected for reference. Satisfaction of this principle is insured by expressing constitutive equations in terms of appropriate tensor quantities.

A constitutive equation should be written in a coordinate system embedded in the material, that is, from the point of view of an observer attached to material points. Of course, measurements are usually made in a stationary coordinate system, and so it is necessary to be able to transform tensors from a "convected" to a stationary coordinate system. The search for appropriate convected transformations, such as in the classic work of Oldroyd (2), is nothing more than the search for appropriate tensor quantities satisfying the principle of material objectivity.

If one wishes to understand modern theoretical studies of continuum mechanics, he must study the algebra and calculus of tensors, and particularly the application of the tensor language to the kinematics of motion in a continuum. In the Appendix a brief introduction to this topic may be found. Because the tensor analysis of continuum mechanics is an integral part of modern research in rheology, some basic kinematic concepts will be discussed in this chapter.

A. Kinematics of Convected Coordinates

A convected coordinate system is a coordinate system to be thought of as embedded in the material, and deforming as the material undergoes its natural response to the forces acting upon it. If X^i is a convected coordinate of some material point P at time t, then X^i is also a convected coordinate of the *same* point for *all* time, since, by definition, the convected coordinate system moves with the material points. Of course one may define *fixed* coordinates x^i for any material point P, in which case the x^i vary with time as the motion proceeds.

As defined in Eq. A.77 of the Appendix, the (squared) distance between two points at time t may be written in terms of the *metric tensor* g of the chosen coordinate system, so that, for a convected coordinate system

$$ds^2 = g_{ij}\, dX^i\, dX^j \tag{3.1}$$

In a general motion referred to convected coordinates it is certainly possible for ds^2 to vary with time, in which case we say that a deformation has occurred. But the X^i, by definition, are time independent. Hence the metric tensor g of a convected coordinate system is a time-dependent metric, the variation of which is some measure of deformation.

Consider now two points of a material whose squared separation at time t is $ds^2(t)$. Suppose at some earlier time t_0 the separation was $ds^2(t_0)$. Then the extent of deformation, or strain, may be expressed as

$$ds^2(t) - ds^2(t_0) = [g_{ij}(t) - g_{ij}(t_0)]\, dX^i\, dX^j \tag{3.2}$$

and the strain tensor **c** may be defined as

$$ds^2(t) - ds^2(t_0) = c_{ij}\, dX^i\, dX^j \tag{3.3}$$

This strain tensor is defined with respect to the *convected* coordinate system. It is referred to the time t_0 and we say that strain has occurred with respect to the configuration at time t_0 if c_{ij} is non-zero. Intuitively one can see that it is possible that the material was already in a state of strain at the reference time t_0 after which no further strain occurred. In that case no strain will exist at time t *relative to the reference state* of the body at time t_0. For this reason one usually chooses the time t_0 to be a time at which the body was in its *natural* (strain free) state.

It will be necessary to consider time rates of change as "seen" by an observer in a convected coordinate system, that is, time rates of change which are natural to the material. In classical hydrodynamics it is common to define a time derivative D/Dt, called the "material" or "substantial" derivative, which represents the time rate of change of a quantity referred to fixed coordinates, the derivative being taken with convected coordinates held constant. The same notation D/Dt, will be maintained for the time rate of change of a quantity referred to *convected* coordinates, the derivative being taken with *convected* coordinates held constant.

Consider, as an important example, the strain tensor **c** (X, t) referred to convected coordinates. The rate of strain tensor will be defined as $D\mathbf{c}/Dt$, the derivative being taken with convected coordinates held constant. Since

CONTINUUM THEORIES

measurements are normally made in a stationary coordinate system, it is necessary to find the components of Dc/Dt referred to fixed coordinates.

Let the components of **c** in a fixed coordinate system be $\gamma_{ij}(\mathbf{x}, t)$. By the rules of tensor transformation

$$c_{mn}(\mathbf{X}, t) = \frac{\partial x^i}{\partial X^m} \frac{\partial x^j}{\partial X^n} \gamma_{ij}(\mathbf{x}, t) \qquad (3.4)$$

and so

$$\frac{Dc_{mn}}{Dt} = \frac{\partial x^i}{\partial X^m} \frac{\partial x^j}{\partial X^n} \frac{D\gamma_{ij}}{Dt} + \frac{\partial x^i}{\partial X^m} \gamma_{ij} \frac{D}{Dt}\left(\frac{\partial x^j}{\partial X^n}\right) + \frac{\partial x^j}{\partial X^n} \gamma_{ij} \frac{D}{Dt}\left(\frac{\partial x^i}{\partial X^m}\right) \qquad (3.5)$$

The second and third terms may be simplified by noting that

$$\frac{D}{Dt}\left(\frac{\partial x^j}{\partial X^n}\right) = \frac{\partial}{\partial X^n}\left(\frac{Dx^j}{Dt}\right) \qquad (3.6)$$

The change in x^j, with convected coordinates held constant, is just the time rate of change of the position of a material particle referred to fixed coordinates, or the velocity component v^j referred to fixed coordinates. Hence Eq. 3.5 becomes

$$\frac{Dc_{mn}}{Dt} = \frac{\partial x^i}{\partial X^m} \frac{\partial x^j}{\partial X^n} \frac{D\gamma_{ij}}{Dt} + \frac{\partial x^i}{\partial X^m} \gamma_{ij} \frac{\partial v^j}{\partial X^n} + \frac{\partial x^j}{\partial X^n} \gamma_{ij} \frac{\partial v^i}{\partial X^m} \qquad (3.7)$$

If v^i is regarded as a function of the x^r, then

$$\frac{\partial v^i}{\partial X^n} = \frac{\partial v^i}{\partial x^r} \frac{\partial x^r}{\partial X^n} \quad \text{and} \quad \frac{\partial v^j}{\partial X^n} = \frac{\partial v^j}{\partial x^r} \frac{\partial x^r}{\partial X^n} \qquad (3.8)$$

This leads to

$$\frac{Dc_{mn}}{Dt} = \frac{\partial x^i}{\partial X^m} \frac{\partial x^j}{\partial X^n} \frac{D\gamma_{ij}}{Dt} + \frac{\partial x^i}{\partial X^m} \frac{\partial x^r}{\partial X^n} \frac{\partial v^j}{\partial x^r} \gamma_{ij} + \frac{\partial x^j}{\partial X^n} \frac{\partial x^r}{\partial X^m} \frac{\partial v^i}{\partial x^r} \gamma_{ij} \qquad (3.9)$$

The second term involves a summation over the indices r and j. Change r to j and j to k. In the third term change r to i and i to k. The result is

$$\frac{Dc_{mn}}{Dt} = \frac{\partial x^i}{\partial X^m} \frac{\partial x^j}{\partial X^n} \left(\frac{D\gamma_{ij}}{Dt} + \gamma_{ik} \frac{\partial v^k}{\partial x^j} + \gamma_{kj} \frac{\partial v^k}{\partial x^i}\right) \qquad (3.10)$$

By inspection of the transformation rule for a tensor (Eq. A.85, Appendix A) it can be seen that

$$\frac{D\gamma_{ij}}{Dt} + \gamma_{ik} \frac{\partial v^k}{\partial x^j} + \gamma_{kj} \frac{\partial v^k}{\partial x^i} \equiv \frac{\delta\gamma_{ij}}{\delta t} \qquad (3.11)$$

must represent the components, referred to fixed coordinates, of the convected derivative Dc_{mn}/Dt. The derivative in *fixed* coordinates,

$D\gamma_{ij}/Dt$, is the usual material derivative given by

$$\frac{D\gamma_{ij}}{Dt} = \frac{\partial \gamma_{ij}}{\partial t} + v^k \frac{\partial \gamma_{ij}}{\partial x^k} \qquad (3.12)$$

The above analysis is true for *any* second-order tensor with components γ_{ij} referred to fixed coordinates. In this particular case γ represents the fixed components of the strain tensor **c**, and $\delta\gamma/\delta t$ is the rate of strain tensor Δ.

That this is so may be seen by writing Eq. 3.11 in terms of covariant derivatives (Appendix, Eq. A.108)

$$\frac{\delta \gamma_{ij}}{\delta t} = \frac{\partial \gamma_{ij}}{\partial t} + v^k \gamma_{ij,k} + v^k_{,i}\gamma_{kj} + v^k_{,j}\gamma_{ik} \qquad (3.13)$$

Upon comparing Eqs. 3.2 and 3.3 it may be seen that

$$\frac{Dc_{ij}}{Dt} = \frac{D\mathcal{g}_{ij}}{Dt} \qquad (3.14)$$

If $D\mathcal{g}_{ij}/Dt$ is regarded as the rate of strain tensor in convected coordinates, then the rate of strain tensor in fixed coordinates is, by Eqs. 3.10 and 3.11, $\delta g_{ij}/\delta t$, where **g** is the metric tensor of the fixed coordinate system. If one uses Eq. 3.13 now, and recalls that **g** is independent of time, and notes further Ricci's Lemma ($g_{ij,k} = 0$),* it follows that

$$\frac{\delta g_{ij}}{\delta t} = g_{kj}v^k_{,i} + g_{ik}v^k_{,j} = v_{j,i} + v_{i,j} = \Delta_{ij} \qquad (3.15)$$

The foregoing analysis is due to Oldroyd (2), and $\delta/\delta t$ is called the Oldroyd derivative. The Oldroyd derivative clearly satisfies the principle of material objectivity since it transforms as a tensor from convected to fixed coordinates. Hence terms like $\delta\gamma_{ij}/\delta t$ may appear in constitutive equations, for any second-order tensor γ.

The significance of the principle of material objectivity may be better understood by considering some quantities which do *not* satisfy the proper transformation laws. For the sake of simplicity this will be done in rectangular coordinates, although the results may be shown to be valid in any orthogonal coordinate system.

Let x'_k and x_j be position coordinates in two rectangular coordinate systems. Allow for the possibility that the two systems may translate and rotate with respect to one another, so that

$$x'_k = a_{kj}x_j + b_k \qquad (3.16)$$

* A tensor equation valid in one coordinate system is valid in any coordinate system. In cartesian coordinates the metric tensor is the Kronecker delta δ. Since the components of δ are constants, it follows that $\delta_{ij,k} = 0$. Hence, in any coordinate system the metric tensor satisfies $g_{ij,k} = 0$.

CONTINUUM THEORIES

where **a** and **b** may be functions of time, but not position. Consider the motion of some continuum, and let

$$v'_k = \frac{Dx'_k}{Dt} \tag{3.17}$$

be the velocity components in the primed frame. Then

$$v'_k = a_{kj}\frac{Dx_j}{Dt} + x_j\frac{\partial a_{kj}}{\partial t} + \frac{\partial b_k}{\partial t} \tag{3.18}$$

or

$$v'_k = a_{kj}v_j + \left\{x_j\frac{\partial a_{kj}}{\partial t} + \frac{\partial b_k}{\partial t}\right\} \tag{3.19}$$

Hence, unless the two reference frames are stationary, so that the bracketed term in Eq. 3.19 vanishes, the velocity vector does not satisfy the principle of material objectivity. This is nothing more than a mathematical expression of the fact that two observers in relative motion ascribe a different velocity to some object under observation.

From Eq. 3.19, and the fact that **a** and **b** are only functions of time, it follows that

$$\frac{\partial v'_k}{\partial x'_m} = a_{kj}\frac{\partial v_j}{\partial x'_m} + \frac{\partial x_j}{\partial x'_m}\frac{\partial a_{kj}}{\partial t} = a_{kj}\frac{\partial v_j}{\partial x_n}\frac{\partial x_n}{\partial x'_m} + \frac{\partial x_j}{\partial x'_m}\frac{\partial a_{kj}}{\partial t} \tag{3.20}$$

As shown in the Appendix (Eq. A.89), terms such as $\partial x_n/\partial x'_m$ are just components a_{nm}. Hence

$$\frac{\partial v'_k}{\partial x'_m} = a_{kj}a_{mn}\frac{\partial v_j}{\partial x_n} + a_{jm}\frac{\partial a_{kj}}{\partial t} \tag{3.21}$$

and it is clear that, in general, the velocity gradient does not transform as a tensor.

By a similar argument it follows that

$$\frac{\partial v'_m}{\partial x'_k} = a_{mj}a_{kn}\frac{\partial v_j}{\partial x_n} + a_{jk}\frac{\partial a_{mj}}{\partial t} \tag{3.22}$$

Note that n and j are dummy indices in the first term on the right-hand side of Eq. 3.22 and may be interchanged. Note also that the **a** tensor is symmetric, so that the subscripts may be interchanged in both terms. The result is

$$\frac{\partial v'_m}{\partial x'_k} = a_{nm}a_{kj}\frac{\partial v_n}{\partial x_j} + a_{kj}\frac{\partial a_{jm}}{\partial t} \tag{3.23}$$

If Eq. 3.23 is added to Eq. 3.21 the result is

$$\frac{\partial v'_k}{\partial x'_m} + \frac{\partial v'_m}{\partial x'_k} = a_{kj}a_{nm}\left(\frac{\partial v_j}{\partial x_n} + \frac{\partial v_n}{\partial x_j}\right) + a_{jm}\frac{\partial a_{kj}}{\partial t} + a_{kj}\frac{\partial a_{jm}}{\partial t}$$

or

$$\Delta'_{km} = a_{kj}a_{nm}\Delta_{jn} + \frac{\partial}{\partial t}(a_{kj}a_{mj}) \tag{3.24}$$

By the orthogonality of the coordinate systems it is known (Appendix, Eq. A.23) that $a_{kj}a_{mj} = \delta_{km}$, and so

$$\frac{\partial}{\partial t}(a_{kj}a_{mj}) = \frac{\partial}{\partial t}\delta_{km} = 0 \tag{3.25}$$

It follows immediately that Δ transforms as a tensor under this general coordinate change, and so Δ satisfies the principle of material objectivity.

If Eq. 3.23 is subtracted from Eq. 3.21 it follows that

$$\frac{\partial v'_k}{\partial x'_m} - \frac{\partial v'_m}{\partial x'_k} = a_{kj}a_{nm}\left(\frac{\partial v_j}{\partial x_n} - \frac{\partial v_n}{\partial x_j}\right) + a_{jm}\frac{\partial a_{kj}}{\partial t} - a_{kj}\frac{\partial a_{jm}}{\partial t} \tag{3.26}$$

If the differentiation of the product on the left-hand side of Eq. 3.25 is carried out, it may be seen that the last two terms on the right-hand side of Eq. 3.26 are equal but of opposite sign. If the vorticity tensor ω is introduced (Appendix, Eq. B.13), it follows that

$$\omega'_{km} = a_{kj}a_{mn}\omega_{jn} - 2a_{kj}\frac{\partial a_{jm}}{\partial t} \tag{3.27}$$

Thus it is seen that vorticity does not transform as a tensor under a time-dependent coordinate transformation, and so it not admissible in a constitutive equation, according to the principle of material objectivity.

Now consider the term

$$a_{kp}\omega'_{km} = a_{kp}a_{kj}a_{mn}\omega_{jn} - 2a_{kp}a_{kj}\frac{\partial a_{jm}}{\partial t}$$

or

$$a_{kp}\omega'_{km} = \delta_{pj}a_{mn}\omega_{jn} - 2\delta_{pj}\frac{\partial a_{jm}}{\partial t} \tag{3.28}$$

If p is replaced by j, and if one now solves for the time derivative, it follows that

$$2\frac{\partial a_{jm}}{\partial t} = a_{mn}\omega_{jn} - a_{kj}\omega'_{km} \tag{3.29}$$

If one of the coordinate frames is regarded as convected and the other as fixed, then this result shows, as expected, that the relative rotation of the convected and fixed frames is related to the vorticity of the flow.

This result may be used to show that the Oldroyd derivative is not the only time derivative of a second rank tensor which obeys the principle of material objectivity. The details are tedious and will not be repeated here (3). It can be demonstrated, for example, that

$$\frac{\mathscr{D}}{\mathscr{D}t} T_{ij} = \frac{\partial T_{ij}}{\partial t} + v_k \frac{\partial T_{ij}}{\partial x_k} - \omega_{jm} T_{mi} - \omega_{im} T_{mj} \qquad (3.30)$$

transforms as a tensor and satisfies the principle of material objectivity. Hence the operator $\mathscr{D}/\mathscr{D}t$ is admissible in a constitutive equation. It is known as the Jaumann derivative, and is different from the Oldroyd derivative $\mathfrak{d}/\mathfrak{d}t$. Other time derivatives appear in the literature (4) which provide a proper transformation from convected to fixed coordinates. It is not possible to choose among these derivatives on theoretical grounds, and there does not seem to be any compelling experimental evidence supporting the use of any single definition of a time derivative.

B. The Simple Fluid

In a study which will without doubt be regarded as of classical importance when histories of mechanics are written in the future, Noll (5) investigated the formulation of constitutive equations by starting with very general and simple notions of the concepts of material and motion. Noll's work is extremely formal and very elegant. It is also quite difficult to study in its original form. An attempt will be made here to present a discussion of Noll's work which will lack some rigor (and most of the elegance) but which will maintain some connection with the ideas which motivated Noll.

If one assumes that the present response of a material is dependent upon the past history of deformation of the material then its constitutive equation should be a relationship between stress and the history of response of the material for all times in the past up to the present time. Furthermore, if one considers some material particle **X**, the only part of the history which affects the response at **X** is the history of an arbitrarily small neighborhood of **X**.

Motion may be described as the transformation of material particles through a succession of places defined with respect to some coordinate system. (Appendix B, pp. 230–237 discusses the kinematics of motion.) For example, a particle **X** may occupy the place $\boldsymbol{\xi}$ at time $t + s$ and the place **x** at the later time t. A measure of motion is the displacement gradient **F**, with mixed components

$$F_j^{i(t)}(\mathbf{X}, s) = \frac{\partial \xi^i(\mathbf{X}, t + s)}{\partial x^j(\mathbf{X}, t)} \qquad (3.31)$$

which are to be thought of as evaluated at the present time t, but dependent upon the time s relative to the present time $(-\infty < s \leq 0)$.

The "history" of the motion at $\mathbf{X}$ is the succession of values taken on by $\boldsymbol{\xi}(\mathbf{X}, t + s)$ for all times s up to and including the present time t $(s = 0)$. One may define a "function of the motion," say $f[\boldsymbol{\xi}(\mathbf{X}, t + s)]$, which would depend upon the value of $\boldsymbol{\xi}$ at any particular time $t + s$. Instead, however, one defines a "functional of the motion," denoted

$$\mathop{\mathfrak{H}}_{s=-\infty}^{s=0} [\boldsymbol{\xi}(\mathbf{X}, t + s)],$$

which depends upon the entire succession of values of $\boldsymbol{\xi}$ evolved by the particle $\mathbf{X}$ as s goes from $-\infty$ to the present time, $s = 0$.

A simple fluid is now defined as one for which the stress at some material point is a functional of the history of the displacement gradients $\mathbf{F}^{(t)}(s)$:

$$\boldsymbol{\tau}(\mathbf{X}, t) = \mathop{\mathfrak{H}}_{s=-\infty}^{s=0} (\mathbf{F}^{(t)}(s)) \tag{3.32}$$

By definition the simple fluid is one for which the functional depends upon the motion only through the first spatial derivatives of $\boldsymbol{\xi}$. The dependence of the stress on past history allows for the possibility of a "memory" of the material for its past configurations, although this need not necessarily be a perfect memory.

The dependence on the displacement gradients $\mathbf{F}^{(t)}(s)$ is usually cast in terms of the Cauchy–Green tensor at time $t + s$ relative to the present time t.* This tensor $\mathbf{C}^{(t)}(s)$ has covariant components, in a fixed coordinate system

$$C_{ij}^{(t)}(s) = g_{kl}(t + s) \frac{\partial \xi^k(t + s)}{\partial x^i(t)} \frac{\partial \xi^l(t + s)}{\partial x^j(t)} \tag{3.33}$$

where $\mathbf{g}(t + s)$ is the metric tensor at the place $\boldsymbol{\xi}(t + s)$.

In order to make these formal definitions less obscure it is helpful to consider a particularly simple type of flow, the so-called "simple shear flow."† Suppose a motion exists such that a particle occupying the place $\boldsymbol{\xi}(t + s)$ at time $t + s$ finds itself at the place $\mathbf{x}(t)$ at time t, with $\boldsymbol{\xi}$ and $\mathbf{x}$ connected by

$$\begin{aligned} \xi^1 &= x^1 + sv^1(x^2) \\ \xi^2 &= x^2 \\ \xi^3 &= x^3 \end{aligned} \tag{3.34}$$

---

* This tensor employs the *present* configuration as the *reference* configuration.
† A definition of the simple shear flow is given in Chapter 1, p. 8.

CONTINUUM THEORIES

This is a particularly simple motion since particles remain in surfaces of constant ξ^2 and ξ^3, and move only in the 1-direction with a constant velocity $v^1(x^2)$. The velocity component v^1 is a function only of the coordinate x^2. The flows pictured in Table 1.6 (p. 9) satisfy these general requirements.

Let the metric of the coordinate system in use be given by

$$g_{ij} = \begin{pmatrix} h_1^2 & 0 & 0 \\ 0 & h_2^2 & 0 \\ 0 & 0 & h_3^2 \end{pmatrix} \tag{3.35}$$

and suppose the coordinate surfaces are such that $h_i = f_i(x^2, x^3)$, but the h_i are not functions of x^1. The coordinate systems normally used (rectangular, polar, cylindrical, spherical) all satisfy this requirement.

An example of such a flow description is given by laminar axial steady shear flow in a long circular pipe. If a cylindrical coordinate system is taken with the z-axis coincident with the pipe axis then the coordinates of a material particle satisfy

$$\begin{aligned} z(t+s) &= z(t) + sv^z(r) \\ r(t+s) &= r(t) \\ \theta(t+s) &= \theta(t) \end{aligned} \tag{3.34a}$$

and the covariant components of the metric tensor are

$$g_{ij} = \begin{pmatrix} 1 & 0 & 0 \\ 0 & r^2 & 0 \\ 0 & 0 & 1 \end{pmatrix} \tag{3.35a}$$

Similar considerations of other flows commonly achieved in viscosity and normal stress experiments show that these flows all belong to the class of simple shear flows.

The covariant components of the Cauchy-Green tensor are easily found from Eqs. 3.33, 3.34, and 3.35, with the result

$$C_{ij} = \begin{pmatrix} h_1^2 & h_1^2 sv' & 0 \\ h_1^2 sv' & h_2^2 + h_1^2 s^2 v'^2 & 0 \\ 0 & 0 & h_3^2 \end{pmatrix} \tag{3.36}$$

where $v' = dv^1/dx^2$.

The *physical* components of **C** are obtained from the rule (see Appendix A, p. 226)

$$C_{(ij)} = C_{ij}/h_i h_j \quad \text{(no summation)} \tag{3.37}$$

with the result

$$C_{(ij)} = \begin{pmatrix} 1 & \dot{\gamma}s & 0 \\ \dot{\gamma}s & 1 + s^2\dot{\gamma}^2 & 0 \\ 0 & 0 & 1 \end{pmatrix} \quad (3.38)$$

Here $\dot{\gamma} = (h_1/h_2)\, dv^1/dx^2$ is the physical component of the velocity gradient, that is, the shear rate.

It is easy to see that one may decompose $\mathbf{C}$ into the sum

$$\mathbf{C} = \boldsymbol{\delta} + s\mathbf{A} + s^2\mathbf{B} \quad (3.39)$$

where $\boldsymbol{\delta}$ is the Kronecker delta and $\mathbf{A}$ and $\mathbf{B}$ are the tensors

$$A_{ij} = \dot{\gamma}\begin{pmatrix} 0 & 1 & 0 \\ 1 & 0 & 0 \\ 0 & 0 & 0 \end{pmatrix} \text{ and } B_{ij} = \dot{\gamma}^2 \begin{pmatrix} 0 & 0 & 0 \\ 0 & 1 & 0 \\ 0 & 0 & 0 \end{pmatrix} \quad (3.40)*$$

Thus it is seen that in a simple shear flow the deformation history is a particularly simple function of past time s, and it is the same function independent of the time t. The response of two simple fluids differs only if the tensors $\mathbf{A}$ and $\mathbf{B}$ differ. These simplifications allow one to state that, in a simple shear flow, the stress reduces to a *function* of $\mathbf{A}$ and $\mathbf{B}$ and is not as complex as a *functional* of $\mathbf{C}$.

The result then is the mathematical statement that

$$\boldsymbol{\tau} = \mathbf{f}(\mathbf{A}, \mathbf{B}) \quad (3.41)$$

It is not difficult to verify that, for a simple shear flow, $\mathbf{A} = \boldsymbol{\Delta}$, the rate of deformation tensor, and $2\mathbf{B} = \delta\boldsymbol{\Delta}/\delta t$, the Oldroyd derivative defined in Eq. 3.13. Thus an alternate form of Eq. 3.41 is

$$\boldsymbol{\tau} = \mathbf{f}\left(\boldsymbol{\Delta}, \frac{\delta\boldsymbol{\Delta}}{\delta t}\right) \quad (3.42)$$

This is known as a constitutive equation of the "rate type," and is essentially the Rivlin-Ericksen fluid (7). Noll's work shows that the Rivlin-Ericksen fluid is identical to the simple fluid in the case of a simple shear flow.

The components of $\mathbf{A}$ and $\mathbf{B}$ are given, in Eq. 3.40, in a coordinate system whose orientation is implicit in Eqs. 3.34. Suppose a new coordinate

---

* In the original work (6) the $\mathbf{B}$ tensor differs from the one given here. The difference is due to our maintenance of the convention that the 1-axis is the flow direction, while the original work takes this to be the 2-axis.

system is used, obtained from the original set of axes by a reflection corresponding to

$$a_{ij} = \begin{pmatrix} 1 & 0 & 0 \\ 0 & 1 & 0 \\ 0 & 0 & -1 \end{pmatrix} \qquad (3.43)$$

Then, through the use of the rule of transformation given in the Appendix (p. 205), it may be verified that in the new coordinate system **A** and **B** have components

$$A'_{ij} = \dot\gamma\begin{pmatrix} 0 & 1 & 0 \\ 1 & 0 & 0 \\ 0 & 0 & 0 \end{pmatrix} = A_{ij} \quad \text{and} \quad B'_{ij} = \dot\gamma^2\begin{pmatrix} 0 & 0 & 0 \\ 0 & 1 & 0 \\ 0 & 0 & 0 \end{pmatrix} = B_{ij} \quad (3.44)$$

Hence **A** and **B** are unchanged by this particular reflection.

The principle of material objectivity requires that a constitutive equation be invariant with respect to coordinate system, or

$$\boldsymbol{\tau}' = \mathbf{f}(\mathbf{A}', \mathbf{B}') \qquad (3.45)$$

with the understanding that the functional dependence in Eq. 3.45 must be the same as in Eq. 3.41. Since $\mathbf{A}' = \mathbf{A}$ and $\mathbf{B}' = \mathbf{B}$ for the particular transformation chosen, it follows that $\boldsymbol{\tau}' = \boldsymbol{\tau}$. But if the components of $\boldsymbol{\tau}$ are transformed with the use of Eq. 3.43, it may be seen that

$$\tau'_{ij} = \begin{pmatrix} \tau_{11} & \tau_{12} & -\tau_{13} \\ \tau_{21} & \tau_{22} & -\tau_{23} \\ -\tau_{31} & -\tau_{32} & \tau_{33} \end{pmatrix} \qquad (3.46)$$

The components of $\boldsymbol{\tau}'$ are the same as the corresponding components of $\boldsymbol{\tau}$, with the exception of the shear components in the 3-direction. The only possibility for satisfying material objectivity lies in the case $\tau'_{13} = -\tau_{13} = -\tau_{31} = 0$, $\tau'_{23} = -\tau_{23} = -\tau_{32} = 0$. Hence, in the simple shear flow of the simple fluid, the stress tensor is

$$\tau_{ij} = \begin{pmatrix} \tau_{11} & \tau_{12} & 0 \\ \tau_{21} & \tau_{22} & 0 \\ 0 & 0 & \tau_{33} \end{pmatrix} \qquad (3.47)$$

These five non-zero components of $\boldsymbol{\tau}$ are not independent but are subject to the symmetry relations $\tau_{12} = \tau_{21}$ and the condition of the vanishing of the sum

$$\tau_{11} + \tau_{22} + \tau_{33} = 0 \qquad (3.48)*$$

* This is equivalent to the assumption that $-p = \frac{1}{3}(T_{11} + T_{22} + T_{33})$.

Hence only three of the stresses are independent, these being the shear stress and two of the normal stresses. Since the stresses depend only on **A** and **B**, and these in turn depend on a single variable, the shear rate $\dot\gamma$, it follows that the stresses are functions only of $\dot\gamma$ in the simple shear flow. This may be expressed by the set of equations (6)

$$\tau_{12} = \tau(\dot\gamma)$$
$$\tau_{11} - \tau_{33} = \sigma_1(\dot\gamma) \qquad (3.49)$$
$$\tau_{22} - \tau_{33} = \sigma_2(\dot\gamma)$$

where the two independent normal stresses have been arbitrarily taken to be the normal stress differences shown in Eq. 3.49. The three functions τ, σ_1, and σ_2 are material functions and are easily recognized to be related to the viscosity and normal stress coefficients defined in Chapter 1 by

$$\tau = \dot\gamma\eta(\dot\gamma)$$
$$\sigma_1 = \dot\gamma^2\psi_{13} \qquad (3.50)$$
$$\sigma_2 = \dot\gamma^2\psi_{23}$$

and

$$\psi_{12} = (\sigma_1 - \sigma_2)/\dot\gamma^2$$

Equations 3.49 establish the fact that three material functions are sufficient to describe a simple shear flow of a simple fluid. Conversely, it is seen that the maximum information that one may obtain from a simple shear flow of a simple fluid is these three functions. One does not expect, nor does he find, that these three material functions are always sufficient to characterize the more general functional dependence of stress on the deformation history of flows more complex than simple shear.

It should be emphasized also that it is certainly possible that the simple fluid concept does not include all real fluids. The value of the simple fluid concept lies in the fact that many specific constitutive equations proposed by various investigators may be shown to be special cases of the simple fluid. Hence the simple fluid model may be used to establish very general results applicable to a wide variety of specific models of response, and thereby establish certain elements of unity among these models.

Coleman and Noll (8), for example, developed the relationship of simple fluid theory to viscoelasticity. Starting with the concept of the simple fluid, they added the idea of a "fading memory." This notion implies that the dependence of stress on the deformation history is weakest for history in the far past and strongest for recent history. Just as a smooth *function* may be expressed as an infinite series of *powers* of its independent variables, so too a *functional* may be expanded into an infinite series of *integrals* of the history upon which it depends. Coleman and Noll give the first term of this expansion as

$$\tau = \int_0^\infty \mu(s) \mathbf{J}(s)\, ds \qquad (3.51)*$$

where $\mathbf{J} = \mathbf{C} - \boldsymbol{\delta}$ and $\mu(s)$ is an unknown function of time. It is a material function which characterizes the memory of the fluid for its past history. Just as power series coefficients must be determined from a knowledge of behavior of the function represented, so $\mu(s)$ must be established through prediction or observation of the memory of the fluid.

An interesting relationship may be developed between $\mu(s)$ and the material functions, defined in Eq. 3.50, which describe a simple shear flow. $\mathbf{J}(s)$ is given explicitly in Eq. 3.39. If the integration indicated in Eq. 3.51 is carried out, one finds

$$\tau = \dot{\gamma} \int_0^\infty s\mu(s)\, ds \quad \text{or} \quad \eta = \int_0^\infty s\mu(s)\, ds \qquad (3.52)$$

and

$$\sigma_1 - \sigma_2 = -\dot{\gamma}^2 \int_0^\infty s^2\mu(s)\, ds \quad \text{or} \quad \psi_{12} = -\int_0^\infty s^2\mu(s)\, ds \qquad (3.53)$$

Since $\mu(s)$ is a function only of time s, and in particular is independent of shear rate, it follows from Eqs. 3.52 and 3.53 that the fluid defined by Eq. 3.51 shows a newtonian viscosity and a constant normal stress coefficient. This is a consequence of the "linearization" of the integral expansion of Eq. 3.32 to the single integral shown in Eq. 3.51. The fluid defined by Eq. 3.51 is called a "first-order viscoelastic fluid."[†]

Since a viscoelastic fluid is generally thought of as one which exhibits stress relaxation, it is instructive to examine a *transient* flow of a first-order viscoelastic fluid. Consider, for example, an experiment in which stress relaxation is observed under constant (shear) strain. Suppose there is no strain during the interval $0 \leqslant t < t_0$. Then, at time t_0, a shear strain of constant magnitude γ is imposed. The representation of this deformation is[‡]

$$\begin{array}{ll} 0 \leqslant t < t_0 & t_0 \leqslant t < \infty \\ \xi_1 = x_1 + \gamma x_2 & \xi_1 = x_1 \\ \xi_2 = x_2 & \xi_2 = x_2 \\ \xi_3 = x_3 & \xi_3 = x_3 \end{array} \qquad (3.54)$$

---

* Note that s has been redefined to range over positive values.
† In fact, the second-order fluid also shows constant viscosity and normal stress coefficients (8).
‡ The *strained* configuration is taken as the *reference* configuration.

From the definition of **C** (Eq. 3.33) one may calculate $\mathbf{J} = \mathbf{C} - \boldsymbol{\delta}$ and find

$$\mathbf{J} = 0 \qquad \text{for} \quad 0 \leqslant s < t - t_0$$

$$\mathbf{J} = \gamma \begin{pmatrix} 0 & 1 & 0 \\ 1 & \gamma & 0 \\ 0 & 0 & 0 \end{pmatrix} \qquad \text{for} \quad t - t_0 \leqslant s < \infty \tag{3.55}$$

If **J** is substituted into Eq. 3.51 it follows that

$$\tau_{12} = \int_{t-t_0}^{\infty} \mu(s)\gamma \, ds \tag{3.56}$$

A (shear stress) relaxation modulus $G(t - t_0)$ may be defined by

$$G(t - t_0) = \tau_{12}/\gamma = \int_{t-t_0}^{\infty} \mu(s) \, ds \tag{3.57}$$

If a new function $\phi(s)$ is defined in such a way that

$$\mu(s) = -d\phi/ds \tag{3.58}$$

then it may be seen that Eq. 3.57 is integrable immediately to give

$$G(t - t_0) = \phi(t - t_0) - \phi(\infty) \tag{3.59}$$

If $\phi(\infty)$ is assumed to be zero* then it is clear that $\phi(t)$ is just the relaxation modulus previously defined as $G(t)$. Thus one finds that the relaxation modulus, the viscosity, and the normal stress coefficients are all related to the memory function $\mu(s)$.

It should be recalled that these results assume the validity of Eq. 3.51 as a constitutive equation. Since only one term of the complete integral expansion of the simple fluid is considered, it is reasonable to suppose that Eq. 3.51 is only an approximation to real behavior. Of course this is evident upon examining data, although the order of the approximation is not apparent, in the sense that one can not say *a priori* what the upper limit of shear strain or shear rate might be for application of the linear viscoelastic model.

In order to predict more realistic response two approaches are possible. In one, higher order integrals, that is, additional terms in the expansion of the general history functional, may be considered. This approach leads to very cumbersome equations. Another possibility is to maintain the simple form of the single integral expansion but to replace $\mu(s)$ by a

---

* This corresponds to the idea that no residual stress would remain in the fluid upon complete relaxation.

memory function which depends upon the stress history (9). In the simple shear flow this has the effect of making $\mu(s)$ a function of shear rate. These approaches will not be discussed further here. Bogue (10,11) has presented a discussion of the ideas presented above and has given an example of a particular modification of the integral representation of the simple fluid which predicts viscosity and normal stress coefficients which are shear rate dependent.

II. SPECIAL CLASSES OF CONSTITUTIVE EQUATIONS

A. The Purely Viscous Fluid

The purely viscous fluid is described by a constitutive equation of the general form

$$\tau = \eta \Delta \tag{3.60}$$

where η is the viscosity of the fluid. It is easy to verify that such a model fails to describe two of the commonly observed phenomena of polymeric fluids: stress relaxation and normal stress development in a simple shear flow. The failure to show stress relaxation is apparent since τ vanishes instantaneously when $\Delta = 0$, that is, upon the removal of deformation. If only the shear components of Δ are non-zero, as in a simple shear flow, then the only non-zero components of τ are shear components also, since η is defined to be a *scalar* quantity. Hence this fluid cannot describe normal stress observations in simple shear.

If η is taken to be a scalar function of the rate of deformation tensor Δ, then it can only be a function of the *scalar invariants* of Δ:*

$$I_\Delta = \Delta_{ii} = 2 \operatorname{div} \mathbf{v}$$
$$II_\Delta = \Delta_{ik}\Delta_{ki} \tag{3.61}$$
$$III_\Delta = \epsilon_{ijk}\Delta_{1i}\Delta_{2j}\Delta_{3k}$$

For an incompressible fluid I_Δ vanishes, and for a simple shear flow it may be verified that III_Δ vanishes and $II_\Delta = 2\dot{\gamma}^2$. In some viscometric flow then one may establish the viscosity dependence on shear rate, given in the invariant notation as

$$\eta = \eta(\sqrt{\tfrac{1}{2}II_\Delta}) \tag{3.62}$$

In non-simple shear flows it is common to maintain Eq. 3.62 even though III_Δ does not vanish. While this is only an approximation, experimental evidence indicates that it may be a reasonable approximation for predicting the shear behavior of non-simple flows of some fluids (12,13).

Since Eq. 3.60 fails to describe stress relaxation and normal stress

---

* See Appendix A, p. 213.

phenomena, one should not expect this simple viscous model to give a good representation of shear behavior if the flow involves significant acceleration (either spatial or in time). Since the purely viscous model is widely used, however, and since it has enjoyed some degree of success, some of the more useful forms of Eq. 3.62 will be described below.

1. The Power Law

In this model the viscosity is given by

$$\eta = K(\sqrt{\tfrac{1}{2}II_\Delta})^{n-1} \qquad (3.63)$$

where K and n are taken to be constants. The power law predicts that τ vs. $\dot{\gamma}$ is a straight line on a double logarithmic plot. Many fluids approximate this behavior over about one decade in shear rate (cf. Fig. 2.2, p. 18) but show significant curvature over wider ranges of shear rate (cf. Fig. 2.5, p. 25). Hence, if the shear rate does not vary widely over a particular flow field the power law may provide an adequate description of shear behavior. Since n is less than unity for most materials, Eq. 3.63 predicts an infinite viscosity in the limit of vanishing shear rate. This, of course, is not observed in real fluids, and is a major drawback to the usefulness of the power law. But the great advantage of the power law, its algebraic simplicity, and the small number of adjustable constants (K and n), overcompensate for its poor behavior at low shear, and so one finds the power law in common use in engineering calculations. Examples of the use of the power law to describe viscous effects in complex flows may be found for turbulent pipe flow (14), power requirements for agitators (15), convection in boundary-layer flows (16), flow through porous media (17), and flow in melt extruders (18).

2. The Ellis Model

The Ellis model uses three parameters and may be written in the form (19)

$$1/\eta = 1/\eta_0[1 + (\tau/\tau_{1/2})^{(1-n)/n}] \qquad (3.64)$$

At low shear rates this model approaches newtonian behavior with a zero shear viscosity η_0. At high shear rates power law behavior is approached with n corresponding to the flow index n of Eq. 3.63. $\tau_{1/2}$ is the shear stress at which the viscosity has fallen to $\tfrac{1}{2}\eta$. In comparison to the power law the Ellis model is slightly more complicated algebraically, and requires the measurement of one additional parameter. It fits data over a wider range of shear rate than does the power law, and does not suffer from the "zero shear failure," the prediction of infinite viscosity at zero shear rate. The Ellis model has been widely used in attempts to describe complex flows of non-newtonian fluids (20–23).

3. The Prandtl-Eyring Fluid (24,25)

The viscosity of this fluid is given by

$$\eta = \frac{\eta_0 \operatorname{arcsinh}(B\sqrt{\tfrac{1}{2}II_\Delta})}{B\sqrt{\tfrac{1}{2}II_\Delta}} \tag{3.65}$$

and can be derived from the Eyring kinetic theory of liquids. At low shear rates newtonian behavior, of viscosity η_0, is approached.

The advantage of this model is its ability to reproduce the general behavior of η with only two constants. Its major disadvantage is its algebraic complexity, which often makes difficult the solution of relatively simple flow problems.

A generalization of Eq. 3.65 is sometimes seen (26), in which η is written as an infinite series of the form

$$\eta = \sum_{i=1}^{\infty} \frac{\eta_{0i} \operatorname{arcsinh}(B_i\sqrt{\tfrac{1}{2}II_\Delta})}{B_i\sqrt{\tfrac{1}{2}II_\Delta}} \tag{3.66}$$

The resultant multiparameter model may be fitted quite well to data, but the solution of simple flow problems is very cumbersome, even if the series is truncated after two or three terms. If one is going to use a four- or six-parameter truncation of Eq. 3.66, he might find that a simple power series in II_Δ, containing the same number of coefficients, would fit the data as well with a savings in algebra.

4. The Powell-Eyring Fluid (27)

Many polymeric fluids exhibit newtonian behavior in the limit of very *high* shear rates, as well as low shear rates (28,29). This level of shear rate is not always achieved in common viscometric experiments, and is often much higher than the shear rates encountered in common technological applications. The "upper newtonian viscosity" η_∞ is often lower than η_0 by many orders of magnitude. For these reasons one is not usually concerned about the failure of the models above to predict a finite value of η_∞. However, some empirical relations have been proposed to describe this behavior.

The Powell-Eyring model (27), for example, gives the apparent viscosity as

$$\eta = \eta_\infty + (\eta_0 - \eta_\infty)\frac{\operatorname{arcsinh} B\sqrt{\tfrac{1}{2}II_\Delta}}{B\sqrt{\tfrac{1}{2}II_\Delta}} \tag{3.67}$$

The defects of this formulation are similar to those of the Prandtl-Eyring fluid.

While many other empirical viscosity expressions have appeared in the literature (13,30–34) none has been studied or used to the extent of the models described above. An interesting review of experimental tests of some of these purely viscous constitutive equations is given by Bird (19).

B. The Viscoelastic Fluid

If one wishes to describe viscoelastic phenomena, such as normal stresses accompanying simple shear flow and stress relaxation in transient flows, then constitutive equations of a different class from the purely viscous fluid need to be examined. Two general classes of equations have met with some success. The first to be discussed may be called "rate equations," as they are constitutive equations in which time rates of change of stress and deformation appear. As long as time rates which satisfy material objectivity are used these equations are equally acceptable, but must stand the test of comparison with real behavior. Differences among these equations are due primarily to the type of time derivative employed. The second class of constitutive equation is the "integral equation," usually a truncation and/or empirical modification of the integral expansion of the simple fluid functional. It will be apparent that not all of these models enjoy equal success in predicting viscoelastic phenomena.

1. Rate Equations

The generalized rate equation may be written in operator form as

$$\mathfrak{P}\tau = \mathfrak{R}\Delta \tag{3.68}$$

where $\mathfrak{P}$ and $\mathfrak{R}$ are time differential operators. This class of fluids includes the classical Maxwell fluid (35)

$$\tau + \lambda_1 \frac{\partial \tau}{\partial t} = \eta_0 \Delta \tag{3.69}$$

as a special case, when $\mathfrak{P} = 1 + \lambda_1 \frac{\partial}{\partial t}$ and $\mathfrak{R} = \eta_0$.

Equation 3.69 is inadmissible on the failure of $\partial/\partial t$ to satisfy material objectivity. If one wishes to ignore this objection, on the grounds that $\partial/\partial t$ is a valid approximation to the convected derivative in the limit of very small velocities and deformations (see, for example, Eq. 3.30), then the fact remains that the Maxwell fluid fails to exhibit normal stresses and non-newtonian flow in a simple shear flow. The Maxwell fluid does exhibit stress relaxation and a dynamic viscosity, however. For a deformation history of the form

$$\Delta_{12} = \dot{\gamma} \quad -\infty < t \leqslant 0$$
$$\Delta_{12} = 0 \quad t > 0$$

the stress relaxes according to

$$\tau_{12} = \eta_0 \dot{\gamma} e^{-t/\lambda_1} \tag{3.70}$$

For a sinusoidal deformation, say

$$\dot{\gamma} = \text{Re}\{\dot{\mathbf{\Gamma}} e^{i\bar{\omega}t}\}$$

it is easy to verify that (see Chapter 2, Eq. 2.157) $\eta^* = \eta_0/(1 + i\bar{\omega}\lambda_1)$, which is equivalent to the two expressions

$$\eta' = \eta_0/(1 + \bar{\omega}^2\lambda_1^2) \quad \text{and} \quad \eta'' = \eta_0\lambda_1\bar{\omega}/(1 + \bar{\omega}^2\lambda_1^2) \tag{3.71}$$

While the general behavior predicted by Eqs. 3.70 and 3.71 is consistent with observation, neither result is a good quantitative model for viscoelastic behavior, even at very low deformation rates, for which $\partial/\partial t$ is a good approximation to the convected derivative.

One generalization of the Maxwell model is motivated by the idea that the total stress in the material is the superposition of individual stresses, each arising from different modes of motion or from the motion of molecular segments of various sizes.* Thus the total stress is taken as

$$\boldsymbol{\tau} = \sum_{p=1}^{\infty} \boldsymbol{\tau}_p \tag{3.72}$$

The individual stresses $\boldsymbol{\tau}_p$ may then be taken as maxwellian, each corresponding to a specific relaxation time λ_p and viscosity η_p:

$$\boldsymbol{\tau}_p + \lambda_p \frac{\partial \boldsymbol{\tau}_p}{\partial t} = \eta_p \boldsymbol{\Delta} \tag{3.73}$$

For this generalized Maxwell model Eqs. 3.70 and 3.71 are replaced by

$$\tau_{12} = \dot{\gamma} \sum_{p=1}^{\infty} \eta_p e^{-t/\lambda_p} \tag{3.74}$$

and

$$\boldsymbol{\eta}^* = \sum_{p=1}^{\infty} \eta_p/(1 + i\bar{\omega}\lambda_p) \tag{3.75}$$

While relaxation and dynamic viscosity data may be represented by these expressions with some accuracy, this is primarily the result of having an infinite number of constants available. Since the generalized Maxwell model also fails to predict non-newtonian viscosity under steady shear and normal stress effects, it is of limited utility.

One may find a number of examples of constitutive equations of the

---

* This is an example of a mechanistic or particulate idea motivating a continuum representation.

form of Eq. 3.68 which predict a non-newtonian viscosity, normal stresses under steady shear, and stress relaxation. Among the more successful of these is a rate type of equation due to Spriggs (36) which is motivated by the generalized Maxwell model, Oldroyd's studies of convected derivatives (37, 38), and the results of certain molecular theories (39, 40). Equation 3.73 is first modified so as to satisfy material objectivity, and takes the form

$$\tau_p + \lambda_p \mathscr{F} \tau_p = \eta_p \Delta \tag{3.76}$$

$\mathscr{F}\tau_p$ is a convected derivative given by Spriggs as

$$\mathscr{F}\tau_{ik} = \frac{\mathscr{D}\tau_{ik}}{\mathscr{D}t} + \frac{1+\varepsilon}{2}(\tau_{ij}\Delta_{jk} + \tau_{jk}\Delta_{ij} + \tfrac{1}{3}\tau_{jm}\Delta_{jm}\delta_{ik}) \tag{3.77}$$

and $\mathscr{D}/\mathscr{D}t$ is the Jaumann derivative (Eq. 3.30). ε is an adjustable constant. Instead of an infinite set of unknown relaxation times λ_p, molecular theory (see Chapter 4, Eq. 4.60) is used to suggest an empirical expression for the λ_p in terms of only two adjustable constants, with the result

$$\lambda_p = \lambda p^{-\alpha} \tag{3.78}$$

Finally, the infinite set of "viscosities" η_p is replaced by a fourth constant η_0 (which turns out to be the zero shear viscosity) by writing

$$\eta_p = \eta_0 \lambda_p / \sum_{p=1}^{\infty} \lambda_p = \eta_0/p^\alpha Z(\alpha) \tag{3.79}$$

where

$$Z(\alpha) = \sum_{p=1}^{\infty} p^{-\alpha} \tag{3.80}$$

While Eqs. 3.76 to 3.79 are motivated by theoretical considerations it is best to view this constitutive equation as a four-parameter (η_0, α, λ, ε) empirical model which satisfies material objectivity, and which must be subjected to experimental test.

From the work of Spriggs the following predictions for material functions may be made:

viscosity in simple shear

$$\eta(\dot{\gamma})/\eta_0 = \frac{1}{Z(\alpha)} \sum_{p=1}^{\infty} \frac{p^\alpha}{p^{2\alpha} + (\lambda c \dot{\gamma})^2} \tag{3.81}$$

normal stresses in simple shear

$$\psi_{12}(\dot{\gamma})/2\lambda\eta_0 = -\frac{1}{Z(\alpha)} \sum_{p=1}^{\infty} \frac{1}{p^{2\alpha} + (\lambda c \dot{\gamma})^2} \tag{3.82}$$

$$\psi_{32}(\dot{\gamma}) = -\frac{\varepsilon}{2}\psi_{12}(\dot{\gamma}) \tag{3.83}$$

dynamic viscosity

$$\eta'(\tilde{\omega})/\eta_0 = \frac{1}{Z(\alpha)} \sum_{p=1}^{\infty} \frac{p^{\alpha}}{p^{2\alpha} + (\lambda\tilde{\omega})^2} \quad (3.84)$$

$$\eta''(\tilde{\omega})/\eta_0 = \frac{\tilde{\omega}\lambda}{Z(\alpha)} \sum_{p=1}^{\infty} \frac{1}{p^{2\alpha} + (\lambda\tilde{\omega})^2} \quad (3.85)$$

elongational viscosity

$$\eta_e(\dot{\epsilon})/\eta_0 = \frac{3}{Z(\alpha)} \sum_{p=1}^{\infty} \frac{1}{p^{\alpha} - (1+\varepsilon)\lambda\dot{\epsilon}} \quad (3.86)$$

The constant c is related to ε by

$$c^2 = \frac{2 - 2\varepsilon - \varepsilon^2}{3}$$

The shear viscosity function given above exhibits newtonian behavior in the limit of low shear rate. It is possible to show (36) that at high shear rate the viscosity behavior becomes power law with $n = 1/\alpha$. For most polymeric fluids, then, α is a number greater than unity.

If Eq. 3.84 is compared to Eq. 3.81 it may be seen that $\eta(\dot{\gamma})$ and $\eta'(\tilde{\omega})$ are similar functions. In fact, the prediction of the Spriggs model is that $\eta(\lambda c\dot{\gamma})$ is exactly the same as $\eta'(\lambda\tilde{\omega})$. This says that the viscosity–shear rate curve is the same as the dynamic viscosity–frequency curve for the same fluid, if shear rate is shifted by the factor c. (For this reason c is called the "shift factor.") Observations of this similarity of η to η' exist in the literature. Experimental difficulties often make it hard to ascribe a value other than unity to c (41), although some studies indicate $c = 0.7$ (42).

The normal stress coefficient ψ_{12} shows zero shear behavior given by

$$-\psi_{12}^0 = 2f(\alpha)\lambda\eta_0$$

where

$$f(\alpha) = \sum_{p=1}^{\infty} p^{-2\alpha}/Z(\alpha) \quad (3.87)$$

For very high shear rates $-\psi_{12}$ decreases with increasing shear rate according to $(\lambda c\dot{\gamma})^{n-2}$, so that ψ_{12} is predicted to show power law behavior at high shear rates. Note that, since $\tau_{22} - \tau_{11} = -\psi_{12}\dot{\gamma}^2$, the normal stress difference is quadratic at low shear rates, but at high shear rates is power law, with a slope equal to n. This behavior is sketched in Fig. 3.1. While existing data (Chapter 5) offer support for the predicted normal stress behavior, insufficient data exist from which one might draw unequivocal conclusions.

The validity of the Weissenberg hypothesis ($\tau_{33} = \tau_{22}$ or $\psi_{32} = 0$) is governed by the parameter ε. If $\varepsilon = 0$ the Weissenberg relation is

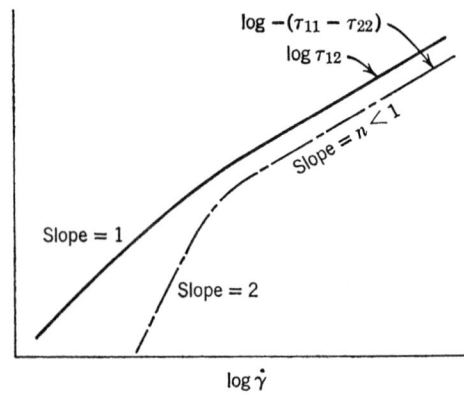

Fig. 3.1. Shear stress and normal stress behavior predicted by Spriggs' model, Eqs. 3.81–3.82.

predicted, and Eq. 3.86 gives $c = 0.816$. Good data from which a test of Eq. 3.86 may be made are lacking, although existing data are not in contradiction.

If Eqs. 3.82 and 3.85 are compared it may be seen that $-\psi_{12}(\dot\gamma)\bar\omega/2\eta_0$ and $\eta''(\bar\omega)/\eta_0$ are the same function, if $c\dot\gamma$ is replaced by $\bar\omega$. Hence two "analogies" are predicted by the Spriggs model, one for η and η' and the other for ψ_{12} and η''. Spriggs presents experimental tests of these analogies and finds fairly good agreement between theory and experiment (36).*

2. The Maxwell Model—Revisited

In this section the Maxwell model is taken up again in terms of, and because of, its impact on the language of rheology. In particular, coefficients and functions whose definitions are motivated by the Maxwell model will be described.

Maxwell's motivation was his recognition that some materials (notably tar) had properties of fluids as well as solids, that is, they exhibited flow phenomena as well as elastic phenomena. The flow response was taken to be newtonian, so that the shear stress τ_{12} was proportional to the shear rate $\dot\gamma_{\text{flow}}$, or

$$\tau_{12} = \eta_0 \dot\gamma_{\text{flow}} \tag{3.88}$$

The elastic behavior was assumed to be hookean, so that the shear strain

---

* Spriggs tests the analogies by plotting $c\dot\gamma(\sigma_1 + \sigma_2)/2\eta_0(1 + \varepsilon)$ vs. η''/η_0 with σ_1 and σ_2 given by Eq. 3.50. Algebraic rearrangement and the substitution of $\bar\omega$ for $c\dot\gamma$ shows this function to be $\bar\omega\psi_{12}/2\eta_0$.

CONTINUUM THEORIES   107

γ_{elastic} was linearly related to the shear stress τ_{12} through a constant, shear modulus G, or

$$\tau_{12} = G\gamma_{\text{elastic}} \tag{3.89}$$

Maxwell assumed that the rate of deformation of a viscoelastic material, due to a given shear stress, would be the *sum* of the rate of deformation due to newtonian flow and the rate of deformation due to elastic response, or

$$\dot{\gamma} = \dot{\gamma}_{\text{flow}} + \frac{d}{dt}\gamma_{\text{elastic}} = \frac{\tau_{12}}{\eta_0} + \frac{1}{G}\frac{d\tau_{12}}{dt} \tag{3.90}$$

One may regard Eq. 3.90 as a differential equation defining the response of a maxwellian material. Solutions may then be obtained for various (simple) stress or strain programs. For example, if the material is subjected to a constant strain γ_0 at time $t \geqslant 0$, the stress relaxes according to

$$\tau_{12} = \tau_0 \exp\left(-tG/\eta_0\right) \tag{3.91}$$

where $\tau_0 = G\gamma_0$ is the initial stress arising from the instantaneous elastic response of the material. If a parameter $\eta_0/G = \lambda$ is defined it is apparent that λ plays the role of a relaxation time in the Maxwell model, and is therefore a measure of how rapidly stress relaxes in a maxwellian material. If the Maxwell model is rewritten in terms of λ and η_0, it takes the form

$$\tau_{12} + \lambda\frac{d\tau_{12}}{dt} = \eta_0\dot{\gamma} \tag{3.92}$$

Stress relaxation at constant strain is usually described in terms of λ and G by writing Eq. 3.91 as

$$G(t) = \tau_{12}(t)/\gamma_0 = G\exp\left(-t/\lambda\right) \tag{3.93}$$

The function $G(t)$ is the "shear relaxation modulus," and should not be confused with the constant shear modulus G. If stress relaxed exponentially G and λ could be obtained by a simple graphical procedure, by plotting the logarithm of $G(t)$ against time and taking the zero time intercept as G and the slope as $\frac{1}{2.3}\lambda$.

But stress relaxation is not usually given by a simple exponential as in Eq. 3.93. The next simplest model which still maintains the maxwellian idea is the sum of exponentials, so that $G(t)$ would be given by

$$G(t) = \sum_{p=1}^{N} G_p \exp\left(-t/\lambda_p\right) \tag{3.94}$$

It is not difficult to show that Eq. 3.94 would follow from a "generalized"

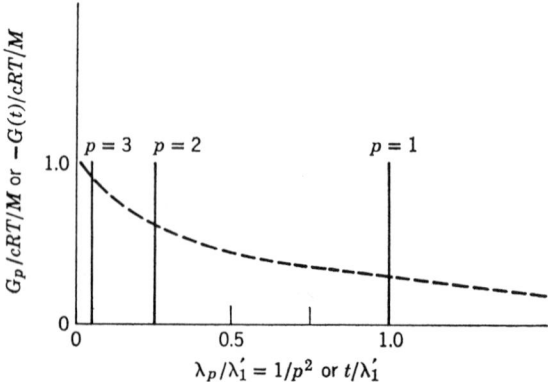

Fig. 3.2. The generalized Maxwell model. Vertical lines represent the first three contributions to the discrete spectrum of shear moduli, according to the Rouse theory. The dashed line gives the corresponding shear relaxation modulus.

Maxwell model, defined by (cf. Eq. 3.73)

$$\tau_p + \lambda_p \frac{d\tau_p}{dt} = \eta_0 \dot{\gamma} \qquad p = 1, \ldots, N \qquad (3.95)$$

$$\tau_{12} = \sum_{p=1}^{N} \tau_p \qquad \eta_0 = \sum_{p=1}^{N} \lambda_p G_p = \sum_{p=1}^{N} \eta_p$$

In the generalized Maxwell model one may think of the set of G_p as a spectrum of shear moduli, each corresponding to a relaxation time λ_p. In general the λ_p constitute a discrete set of numbers, although there may be infinitely many of them (if $N = \infty$). An example of a discrete spectrum of relaxation times is given by the Rouse theory (Eq. 4.60), $\lambda_p = \lambda_1'/p^2$. If all of the G_p are equal, and equal to cRT/M, then complete correspondence to the Rouse theory is attained. Figure 3.2 shows the Rouse $G_p(\lambda_p)$ spectrum, along with the corresponding $G(t)$ curve.

Similar considerations apply to the *tensile* moduli E_p, defined by assuming that the tensile stress in constant tensile strain (that is, constant elongation) e_0 relaxes according to

$$E_r(t) = T_{11}(t)/e_0 = \sum_{p=1}^{N} E_p \exp(-t/\lambda_p) \qquad (3.96)$$

In the region of linear viscoelastic behavior, and for incompressible materials, it may be found that $E_r(t) = 3G(t)$ (43).

Tobolsky and Murakami (44) illustrate the applicability of the generalized Maxwell model to stress relaxation at constant elongation. They describe what they refer to as "procedure X" for obtaining the tensile moduli and relaxation times which appear in Eq. 3.96. The idea of this procedure is quite simple. For sufficiently long times the dominant contribution to $E_r(t)$ is from the term with the maximum relaxation time, or

$$E_r(t) = E_1 \exp(-t/\lambda_1) \quad \text{for large } t \tag{3.97}$$

Hence a plot of log $E_r(t)$ vs. t should be a straight line (for large t) whose slope is $-\frac{1}{2.3}\lambda_1$ and whose zero time intercept is log E_1. With these values established one now repeats the procedure, plotting

$$\log[E_r(t) - E_1 \exp(-t/\lambda_1)] \text{ vs. } t$$

For long times the dominant contribution to this curve would be from the term containing the second largest relaxation time, $E_2 \exp(-t/\lambda_2)$. Hence E_2 and λ_2 are obtained graphically, and the procedure continues in search of the next largest relaxation time.

Tobolsky and Murakami believe that "procedure X" gives accurate estimates of the first two relaxation times, and the corresponding moduli. Beyond this point the graphical procedure is inherently inaccurate since $E_r(t) - \sum_{p=1}^{2} E_p \exp(-t/\lambda_p)$ represents the subtraction of two terms which are large in comparison to their difference. This is simply a reflection of the fact that, except for very short times, the relaxation is dominated by the first two terms of the generalized Maxwell model, in those materials which may be reasonably represented by a generalized Maxwell model.

The form of stress relaxation in the generalized Maxwell model has led to an alternate representation of stress relaxation behavior. The *constants* E_p and λ_p are replaced by a continuous function $E(\lambda)$, defined in such a way that

$$E_r(t) = \int_0^\infty E(\lambda) \exp(-t/\lambda) \, d\lambda \tag{3.98}$$

λ is now regarded as a continuously varying time parameter, and $E(\lambda) \, d\lambda$ represents the contribution to the tensile modulus associated with relaxation times in the range λ to $\lambda + d\lambda$. $E(\lambda)$ is called the "distribution of relaxation times," although it really is a distribution of moduli.

It should not be thought that the integral form of $E_r(t)$ is equivalent to the generalized Maxwell model. Equation 3.98 is a completely different mathematical model for stress relaxation, and is related to Eq. 3.96 only through superficial form and motivation. Equation 3.98 is identical to the generalized Maxwell model *only* in the special case that $E(\lambda)$ is a sum of Dirac delta functions (45)

$$E(\lambda) = \sum_{p=1}^{N} \frac{1}{\lambda^2} E_p \delta\left(\frac{1}{\lambda} - \frac{1}{\lambda_p}\right) \tag{3.99}$$

The Dirac function has the property that $\delta(x - a)$ is zero for all $x \neq a$, and that

$$\int_0^\infty e^{-x} \delta(x - a) \, dx = e^{-a}$$

The form of $E(\lambda)$ for the Maxwell model follows from this property of the Dirac function, and by comparison to Eq. 3.96. A sum of Dirac functions, of course, gives a discrete distribution for $E(\lambda)$, since $E(\lambda)$ is zero everywhere except at the distinct points $\lambda = \lambda_p$.

Because of the common use of a logarithmic time scale in the graphical presentation of relaxation data, $E(\lambda)$ is sometimes replaced by a function $H(\lambda) = \lambda E(\lambda)$ so that

$$E_r(t) = \int_{-\infty}^{\infty} H(\lambda) \exp(-t/\lambda) \, d\ln \lambda \qquad (3.100)$$

If fairly simple functions are *assumed* for $H(\lambda)$ it is possible to integrate Eq. 3.100 and calculate the relaxation response $E_r(t)$ of the material defined by the particular choice of $H(\lambda)$.

It is more difficult to obtain $H(\lambda)$ from $E_r(t)$ rather than the reverse, since Eq. 3.100 would have to be inverted. Actually Eq. 3.100 is similar in form to the relationship between a function and its Laplace transform and Eq. 3.98 is identical to a Laplace transformation between $E_r(t)$ and $\lambda^2 E(\lambda)$*. Since extensive tables and analytical procedures exist for the inversion of the Laplace transform one might hope to be able to obtain $E(\lambda)$, and, using Eq. 3.100, $H(\lambda)$, directly by inversion of the function $E_r(t)$. Unfortunately $E_r(t)$ is not usually of a form which can be inverted by analytical means, and so one is forced to use approximate methods of inversion.

A simple first-order approximation for inversion was proposed by Alfrey (46). In Eq. 3.98 the term $\exp(-t/\lambda)$ is set equal to unity for $\lambda > t$, and zero otherwise. The integral is then, approximately

$$E_r(t) = \int_t^\infty E(\lambda) \, d\lambda \qquad (3.101)$$

Differentiation of this integral gives

$$[dE_r(t)/dt]_{t=\lambda} = -E(\lambda) \qquad (3.102)$$

and the first approximation to $E(\lambda)$ is defined as

$$E_1(\lambda) = -[dE_r(t)/dt]_{t=\lambda} \qquad (3.103)$$

* The Laplace transform of a function $F(s)$ is defined as

$$\mathscr{L}\{F(s)\} = f(t) = \int_0^\infty F(s) e^{-ts} \, ds$$

and the inverse is denoted $F(s) = \mathscr{L}^{-1}\{f(t)\}$. If s is replaced by $1/\lambda$ and $F(s)$ by $\lambda^2 E(\lambda)$, this can be written as $f(t) = \int_0^\infty E(\lambda) e^{-t/\lambda} \, d\lambda$, which is Eq. 3.98.

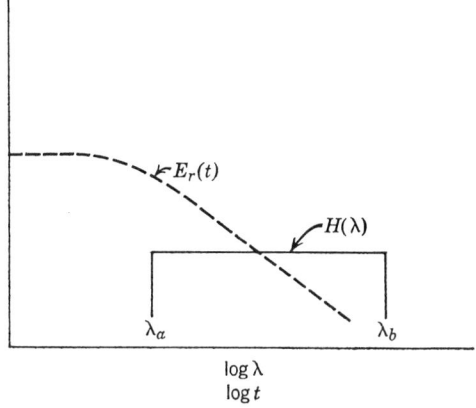

Fig. 3.3. The "box" distribution, and the corresponding relaxation modulus.

In terms of $H(\lambda)$ the first approximation is

$$H_1(\lambda) = \lambda E_1(\lambda) = -[dE_r(t)/d \ln t]_{t=\lambda} = -E_r(t)\left[\frac{d \log E_r(t)}{d \log t}\right]_{t=\lambda} \quad (3.104)$$

Ferry and Williams give a second approximation to $H(\lambda)$ which can be written as (47)

$$H_2(\lambda) = H_1(\lambda)/\Gamma(1 + m) \quad (3.105)$$

where

$$m = -d \log H_1(\lambda)/d \log \lambda \quad (3.106)$$

and $\Gamma(\ )$ is the gamma function (48). Since these approximation methods can require second derivatives of the experimental $E_r(t)$ curve, it is clear that a considerable premium is to be gained by good experimental data. In view of the computational difficulties involved in obtaining $H(\lambda)$ it is not surprising that a fair amount of effort has gone into investigating simple empirical choices of $H(\lambda)$ which lead to realistic relaxation curves $E_r(t)$.

Figure 3.3 shows an example of the type of relaxation response which would be predicted from a "box" distribution

$$\begin{aligned} H(\lambda) &= E_0 & \lambda_a \leq \lambda \leq \lambda_b \\ H(\lambda) &= 0 & \text{for all other } \lambda \end{aligned} \quad (3.107)$$

For this distribution $E_r(t)$ is found to be (49)

$$E_r(t) = E_0[Ei(-t/\lambda_a) - Ei(-t/\lambda_b)] \quad (3.108)$$

where $Ei(\ )$ is the exponential integral function,* tabulated in Ref. 45. Another useful choice for $H(\lambda)$ is the "box-wedge" distribution (50), given by

$$\text{Wedge:} \quad H(\lambda) = E_w/\lambda^{1/2} \quad \lambda_c \leqslant \lambda \leqslant \lambda_d$$
$$\text{Box:} \quad H(\lambda) = E_0 \quad \lambda_a \leqslant \lambda \leqslant \lambda_b \quad (3.109)$$
$$H(\lambda) = 0 \quad \text{for all other } \lambda$$

The wedge distribution *alone* gives rise to relaxation expressed by (43, p. 128)

$$E_r(t) = E_w t^{-1/2}[\Gamma_{t/\lambda_c}(\tfrac{1}{2}) - \Gamma_{t/\lambda_d}(\tfrac{1}{2})] \quad (3.110)$$

where $\Gamma_{t/\lambda}(\tfrac{1}{2})$ is the incomplete gamma function (48).† $E_r(t)$ for the box-wedge distribution is obtained by simply adding Eqs. 3.108 and 3.110. Figure 3.4 shows the relaxation curve corresponding to a box-wedge distribution. Tobolsky (43, pp. 123–133) presents a number of such examples of simple distribution functions used to model real stress relaxation curves.

It should be apparent that the determination of $H(\lambda)$ is of no value in *predicting* stress relaxation if one obtains $H(\lambda)$ from $E_r(t)$ itself. One finds, however, that material response other than stress relaxation may be related to $H(\lambda)$. Hence one has the possibility of using stress relaxation data for the prediction of other types of material response. As an example, consider

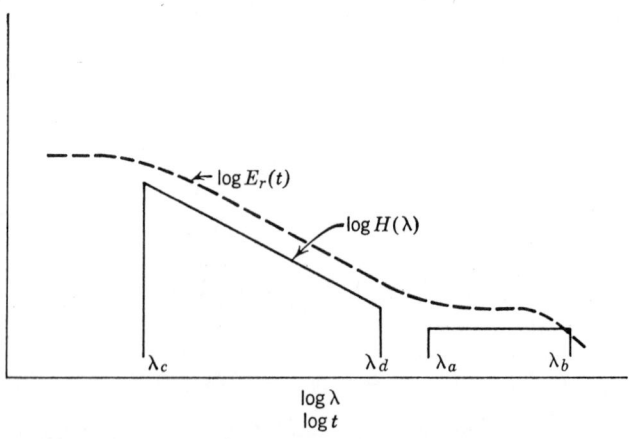

Fig. 3.4. The "wedge-box" distribution, and the corresponding relaxation modulus.

* $Ei(x) \equiv \int_{-\infty}^{x} e^v/v \, dv$.
† $\Gamma_{t/\lambda}(\tfrac{1}{2}) = \int_0^{t/\lambda} e^{-v} v^{-1/2} \, dv$.

the components of the complex viscosity $\boldsymbol{\eta}^*$. The generalized Maxwell model gives†

$$\eta'(\tilde{\omega}) = \sum_{p=1}^{N} \frac{G_p \lambda_p}{1 + \tilde{\omega}^2 \lambda_p^2} \qquad (3.111)$$

and

$$\eta''(\tilde{\omega}) = \sum_{p=1}^{N} \frac{G_p \tilde{\omega} \lambda_p^2}{1 + \tilde{\omega}^2 \lambda_p^2} \qquad (3.112)$$

If a continuous *shear* relaxation modulus $G(\lambda)$ is now defined it is possible to express η' and η'' as

$$\eta'(\tilde{\omega}) = \int_0^\infty \frac{1}{1 + \tilde{\omega}^2 \lambda^2} G(\lambda)\, d\lambda \qquad (3.113)$$

$$\eta''(\tilde{\omega}) = \int_0^\infty \frac{\tilde{\omega}\lambda}{1 + \tilde{\omega}^2 \lambda^2} G(\lambda)\, d\lambda \qquad (3.114)$$

If one had shear stress relaxation data, at constant shear strain, it would be possible to predict the dynamic viscosity behavior of the material. Consideration of continuum theories based on the generalized Maxwell model (such as the Spriggs model discussed on p. 104) indicates that the components of $\boldsymbol{\eta}^*$ may be used to predict the shear viscosity $\eta(\dot{\gamma})$ and the normal stress coefficients $\psi_{ij}(\dot{\gamma})$. Hence it is seen that, in addition to the interrelationships among η, ψ_{ij}, and $\boldsymbol{\eta}^*$, one might also expect (or at least hope for) the *possibility* of predicting these material functions from *stress relaxation* behavior.‡

The most comprehensive test of these ideas is given by Catsiff and Tobolsky (51; 43, pp. 155–156), who took tensile stress relaxation data, giving $E_r(t)$, and predicted the frequency dependence of the components of $\mathbf{G}^* = i\tilde{\omega}\boldsymbol{\eta}^*$. In order to do this it was first necessary to compute the distribution function $H(\lambda)$, with the use of Eqs. 3.105 and 3.106. Then the components of $\mathbf{E}^*$, the dynamic tensile modulus, were calculated from the approximations (43, p. 122)

$$E'(1/\tilde{\omega}) = [E_r(t)]_{t=1/\tilde{\omega}} + [\gamma(m) H_2(\lambda)]_{\lambda=1/\tilde{\omega}} \qquad (3.115)$$

and

$$E''(1/\tilde{\omega}) = [\tfrac{1}{2}\pi(\sec \tfrac{1}{2}m\pi) H_2(\lambda)]_{\lambda=1/\tilde{\omega}} \qquad (3.116)$$

The parameter m is defined in Eq. 3.106, and the function $\gamma(m)$ is tabulated by Catsiff and Tobolsky from

$$\gamma(m) = \tfrac{1}{2}\pi \csc \tfrac{1}{2}m\pi - \Gamma(m) \qquad (3.117)$$

† This follows from Eq. 3.75, with $\eta_p = G_p \lambda_p$.
‡ Recall the simple fluid results, given in Eqs. 3.52 and 3.53.

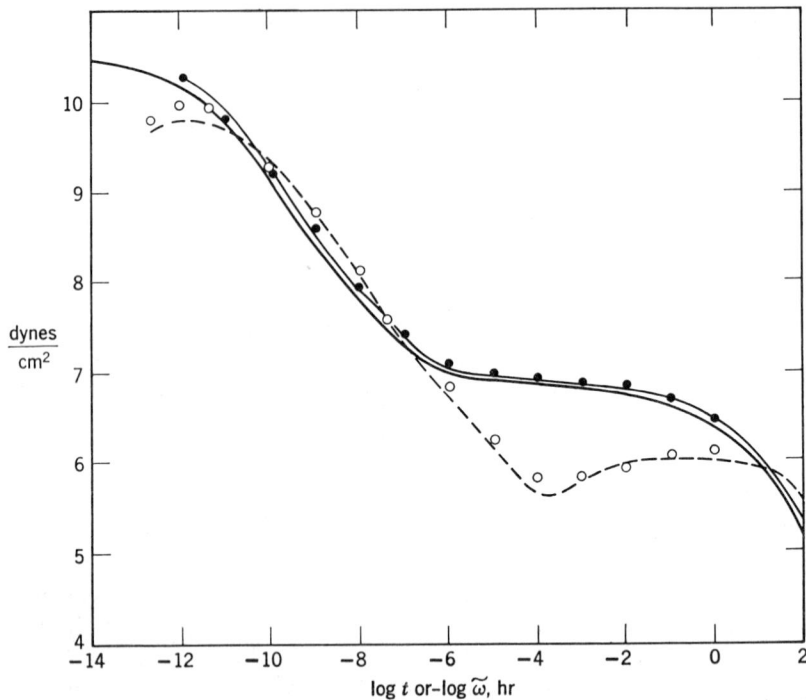

Fig. 3.5. The Catsiff-Tobolsky test of the prediction of the components of **G*** from (———) $E_r(t)$ data (51), (———) $3G'(\tilde{\omega})$ predicted, and (●) measured, (· – – – –) $3G''(\tilde{\omega})$ predicted, and (○) measured. Drawn from Table IV.2 of Ref. 43.

Finally, within the spirit of linear viscoelasticity, and incompressibility, it was assumed that $\mathbf{G}^* = \tfrac{1}{3}\mathbf{E}^*$. Figure 3.5 shows the results of this test for polyisobutylene. There is no question that the procedures outlined here are useful in interrelating different types of rheological response in the materials studied.

3. Integral Representations

In discussing the "simple fluid" concept of Noll, it was noted that the functional of the deformation history could be expanded into an infinite series of integrals, the first term of which is given as Eq. 3.51. Because the complete integral expansion contains an infinite set of unknown functions [of which $\mu(s)$ in Eq. 3.51 is one] such a general formulation is of limited value as a working constitutive equation, and one seeks special cases of these integral representations which contain a finite (and hopefully small) number of unknown material functions.

Actually, some of the simpler integral representations trace their origin back to the Boltzmann superposition principle, formulated in 1874 (52). The Boltzmann principle begins with the notion of a linearly elastic material subject to infinitesimal strain at time t_0. Such a material responds with an infinitesimal stress given by

$$d\tau_{ij} = G\, d\gamma_{ij} \qquad t > t_0 \tag{3.118}$$

Viscoelasticity is then introduced by replacing the shear modulus G by a relaxation function $\phi(t - t_0)$, so that

$$d\tau_{ij}(t) = \phi(t - t_0)\, d\gamma_{ij}(t_0) \tag{3.119}$$

Eq. 3.119 gives a stress which relaxes according to the time $t - t_0$ which has elapsed since the imposition of the strain $d\gamma_{ij}$ at time t_0.

One may then conceive of a material having been subjected to a *sequence* of infinitesimal strains $d\gamma_{ij}^{(n)}$ at times t_n. By *assuming* that the infinitesimal stresses resulting from each strain increment simply add in a linear fashion, with each strain weighted by a relaxation function reckoned from the time of imposition of that strain, then the stress at time t may be written as

$$\tau_{ij}(t) = \sum_{n=0} \phi(t - t_n)\, d\gamma_{ij}(t_n) \tag{3.120}$$

If the sequence of infinitesimal strains is replaced by a continuous strain history, $\gamma_{ij}(t')$, the resultant stress may be written as an integral with the result

$$\tau_{ij}(t) = \int_{-\infty}^{t} \phi(t - t')\, d\gamma_{ij}(t') \tag{3.121}*$$

This linear superposition model, known as the Boltzmann superposition principle, is expected to be valid only for infinitesimal displacements from equilibrium. Within the spirit of this approximation, one may introduce the rate of deformation tensor as $\Delta = d\gamma/dt'$ and rewrite Eq. 3.121 as

$$\tau_{ij}(t) = \int_{-\infty}^{t} \phi(t - t')\, \Delta_{ij}(t')\, dt' \tag{3.122}$$

An alternate formulation may be obtained in terms of strain by integrating Eq. 3.121 by parts with the result

$$\tau_{ij}(t) = \phi(0)\gamma_{ij}(t) + \int_{-\infty}^{t} \mu(t - t')\gamma_{ij}(t')\, dt' \tag{3.123}$$

---

* The integration variable is taken to range from the present time t back to a "beginning" at $t' = -\infty$. This is a matter of convenience, so that one may formulate problems in which a material has been at some steady condition for "all past time," $-\infty < t' < 0$, and is then subjected at $t' = 0$ to a change of deformation.

Here a new function $\mu(t - t')$ is defined which is related to the relaxation function $\phi(t - t')$ by

$$-\frac{d\phi(t - t')}{dt'} = \mu(t - t') \tag{3.124}$$

Written in terms of strain, this form of the Boltzmann superposition principle shows explicitly the elastic nature of the material through the term $\phi(0)\gamma_{ij}(t)$. Suppose, for example, that the material were subjected to an infinitesimal strain history given by

$$\gamma(t') = 0 \quad t' < 0 \tag{3.125}$$
$$\gamma(t') = \gamma_0 \quad t' \geq 0$$

Suppose further that the material relaxed according to

$$\mu(t - t') = \mu_0 e^{-(t-t')/b_t} \tag{3.126}$$

where b_t is some characteristic relaxation time of the material. Then Eq. 3.123 would give

$$\tau_{ij}(t) = \phi(0)\gamma_0 + \gamma_0\mu_0 b_t[e^{-t/b_t} - 1] \tag{3.127}$$

The term $\phi(0)\gamma_0$ is seen to be the instantaneous elastic response of the material. In a *perfectly* elastic material, the relaxation time b_t would approach zero and so τ_{ij} would be, in the limit of $b_t = 0$

$$\tau_{ij}(t) = \phi(0)\gamma_0 = G\gamma_0 \quad t \geq 0 \tag{3.128}$$

which is the expected result.

If one looks back now at the discussion of the single integral truncation of the Coleman-Noll theory, starting with Eq. 3.51, the relationship of the Boltzmann superposition principle to that theory is apparent. The differences between the two formulations disappear if one introduces $s = t - t'$ into the Boltzmann formulation, and if one replaces the strain tensor γ by the strain tensor **J**. In general, one may develop various nonlinear analogs of the Boltzmann model by one or a combination of three devices.* Strain measures other than **J** may be introduced for γ in Eq. 3.123. The function $\mu(t - t')$ may be taken to be some assumed function of the invariants of **J** or **Δ** (10). Finally, the integral expression explicit in shear rate, Eq. 3.122, may be used, but written in the *convected* coordinate system. The act of transforming to the fixed coordinate system introduces certain new features into the model. In particular, such a model exhibits normal stresses in a simple shear flow, as discussed by Oldroyd (2) and by Fredrickson (9). However, as Fredrickson notes, the model

---

* It is not necessary to begin with the Boltzmann principle; it is useful merely as a conceptual bridge to the more complicated expressions.

will show nonlinear viscosity only if the relaxation function is taken to be a function of the rate of deformation.

Recent surveys of integral representations have been presented by Bogue and Doughty (11) and by Spriggs, Huppler, and Bird (53). As yet it is not apparent that constitutive equations written as integral representations offer any better description of real material response than rate equations. The major advantage of integral theories is that they are explicit in stress; if the deformation history is given, the stress follows immediately upon an integration of the form

$$\text{stress} = \int_{-\infty}^{t} (\text{relaxation function})(\text{strain history}) \, dt'$$

Such a procedure is generally simpler to perform than that which would be required of a rate theory, the more realistic of which often require solution of simultaneous differential equations.

For the sake of illustration, an integral representation due to Bird and coworkers (53) will be used to predict the transient response of a material to the sudden application of simple shear. The general form of the constitutive equation is

$$\boldsymbol{\tau} = \int_{-\infty}^{t} \mu[t - t', II_\Delta(t')] \left[\left(1 + \frac{\varepsilon}{2}\right) \mathbf{C}^{-1} + \left(\frac{\varepsilon}{2}\right) \mathbf{C} \right] dt' \quad (3.129)$$

The relaxation function is taken to depend on the "elapsed time" $t - t'$, and on the rate of deformation, through the second invariant II_Δ. ε is an empirical parameter. Bird notes that if μ is taken to be

$$\mu = \left(\eta_0 / \sum_{p=1}^{\infty} \lambda_p \right) \sum_{p=1}^{\infty} \frac{1}{\lambda_p} \frac{1}{1 + \frac{1}{2} II_\Delta(t') c^2 \lambda_p^2} e^{-(t-t')/\lambda_p} \quad (3.130)$$

with $\lambda_p = \lambda/p^\alpha$ then the same results are obtained as with the Spriggs model (p. 104), in the cases of steady shear flow, oscillatory shear flow, and stress relaxation after steady shear flow, provided that c and ε are connected, as in the case of the Spriggs model, by Eq. 3.86. $\mathbf{C}^{-1}$ is the inverse of the Cauchy-Green tensor $\mathbf{C}$, and is called the Finger strain tensor.

Consider a simple shearing flow, for which the physical components of $\mathbf{C}$ are (Eq. 3.38)

$$\mathbf{C} = \begin{pmatrix} 1 & \dot{\gamma}s & 0 \\ \dot{\gamma}s & 1 + s^2\dot{\gamma}^2 & 0 \\ 0 & 0 & 1 \end{pmatrix} \quad \text{for} \quad t' > 0 \quad (3.131)$$

where $s = -(t - t')$. Since, by definition of the inverse of a tensor

$\mathbf{CC}^{-1} = \boldsymbol{\delta}$, it is simple to find that

$$\mathbf{C}^{-1} = \begin{pmatrix} 1 + s^2\dot{\gamma}^2 & -\dot{\gamma}s & 0 \\ -\dot{\gamma}s & 1 & 0 \\ 0 & 0 & 1 \end{pmatrix} \quad \text{for } t' > 0 \quad (3.132)$$

At the current instant of time $t' = t$, $s = 0$, and it follows that $\mathbf{C} = \mathbf{C}^{-1} = \boldsymbol{\delta}$. This reflects the fact that the strain tensors are being measured relative to the current (changing) configuration, rather than relative to the initial configuration. If it is assumed that the material is without deformation for $t' \leq 0$, then $\mathbf{C}$ is the same for all $t' \leq 0$. At $t' = 0$, for which $s = -t'$ it may be seen that

$$\mathbf{C} = \begin{pmatrix} 1 & -\dot{\gamma}t & 0 \\ -\dot{\gamma}t & 1 + \dot{\gamma}^2 t^2 & 0 \\ 0 & 0 & 1 \end{pmatrix} \quad (3.133)$$

while $\mathbf{C}^{-1}$ is found to be

$$\mathbf{C}^{-1} = \begin{pmatrix} 1 + \dot{\gamma}^2 t^2 & \dot{\gamma}t & 0 \\ \dot{\gamma}t & 1 & 0 \\ 0 & 0 & 1 \end{pmatrix} \quad (3.134)$$

These latter two expressions hold, then, for all $t' \leq 0$. The shear stress τ_{12} is found to approach equilibrium according to

$$\tau_{12} = \left(\eta_0 \Big/ \sum_{p=1}^{\infty} \lambda_p\right) \sum_{p=1}^{\infty} \frac{\lambda_p \dot{\gamma}}{[1 + (c\dot{\gamma}\lambda_p)^2]} \left\{1 - \left[1 - (c\dot{\gamma}\lambda_p)^2 \frac{t}{\lambda_p}\right] e^{-t/\lambda_p}\right\} \quad (3.135)$$

At steady state ($t = \infty$), the shear stress is

$$\tau_{12} = \sum_{p=0}^{\infty} \frac{\eta_0 \dot{\gamma} \lambda_p / \sum \lambda_p}{1 + \dot{\gamma}^2 c^2 \lambda_p^2} \quad (3.136)$$

from which it follows that the apparent viscosity is

$$\eta = \tau_{12}/\dot{\gamma} = \sum_{p=0}^{\infty} \frac{\eta_0 \lambda_p / \sum \lambda_p}{1 + \dot{\gamma}^2 c^2 \lambda_p^2} \quad (3.137)$$

Equation 3.137 is identical with Eq. 3.81.

The normal stress difference $\tau_{11} - \tau_{22}$ is found to be a function of time given by

$$\tau_{11} - \tau_{22} = \left(\eta_0 \Big/ \sum_{p=1}^{\infty} \lambda_p\right) \sum_{p=1}^{\infty} \frac{2(\lambda_p \dot{\gamma})^2}{[1 + (c\dot{\gamma}\lambda_p)^2]}$$

$$\times \left\{1 - \left[1 + \frac{t}{\lambda_p} - (c\dot{\gamma}\lambda_p)^2 \frac{t}{2\lambda_p}\right] e^{-t/\lambda_p}\right\} \quad (3.138)$$

At equilibrium the normal stress coefficient ψ_{12} is given by Eq. 3.82.

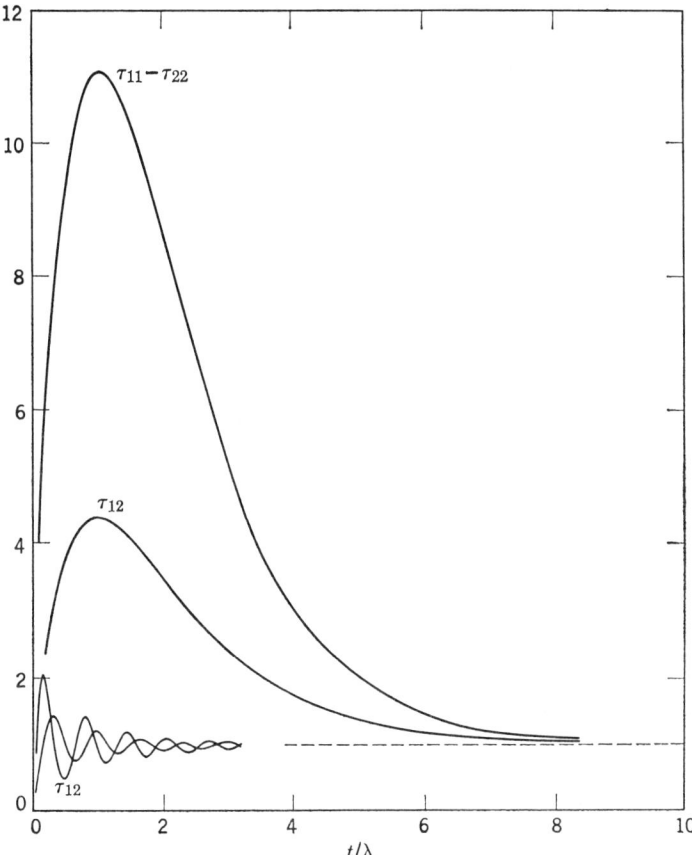

Fig. 3.6. The transient response to sudden imposition of a simple shear flow, according to Eqs. 3.135 and 3.138. Also shown is the oscillatory response of a material obeying Spriggs' rate equation, Eqs. 3.76–3.80, in the lower pair of curves. The curves are normalized to their respective equilibrium values, and are drawn for the case $\alpha = 3$, $c\lambda\dot{\gamma} = 10$.

Figure 3.6 shows the transient approach of τ_{12} and $\tau_{11} - \tau_{22}$ to their steady state values. Huppler (54) has noted that while the steady state results predicted by this integral equation are identical with those of the Spriggs rate equation discussed earlier, the transient approach to equilibrium is different for these two models. Experimental results show an "overshoot" of the equilibrium values, but no oscillation, and neither model is in quantitative agreement with present experimental data on this effect.

Other examples of integral constitutive equations, including solutions of specific flow problems, may be found in the work of Fredrickson (9),

Bogue (10), and Zapas and Craft (55). The latter authors make use of an elastic fluid integral theory developed by Bernstein, Kearsley, and Zapas (56). This so-called BKZ theory takes the form

$$\tau = \int_{-\infty}^{t} [\mu_1(t - t')\mathbf{C} + \mu_2(t - t')\mathbf{C}^{-1}] \, dt' \qquad (3.139)^*$$

where μ_1 and μ_2 are unspecified scalar functions of time, and possibly of the invariants of $\mathbf{C}$ and $\mathbf{C}^{-1}$, and play the role of relaxation functions. The deformation tensors $\mathbf{C}$ and $\mathbf{C}^{-1}$ are taken with respect to the *present time* as the reference configuration. In the BKZ theory μ_1 and μ_2 are not independent, but are related to an "elastic potential" U by

$$\mu_1 = -2 \, \partial U/\partial II_{\mathbf{C}^{-1}} \quad \text{and} \quad \mu_2 = 2 \, \partial U/\partial I_{\mathbf{C}^{-1}} \qquad (3.140)$$

U depends upon the invariants $I_{\mathbf{C}^{-1}}$ and $II_{\mathbf{C}^{-1}}$, as well as the elapsed time $t - t'$. The motivation for introducing the function U lies in the theory of perfectly elastic materials, and will be discussed subsequently.

Zapas and Craft examine behavior under simple elongation, for which see Appendix B, p. 237.

$$\mathbf{C} = \begin{pmatrix} \alpha_1^2 & 0 & 0 \\ 0 & \alpha_2^2 & 0 \\ 0 & 0 & \alpha_3^2 \end{pmatrix} \quad \text{and} \quad \mathbf{C}^{-1} = \begin{pmatrix} 1/\alpha_1^2 & 0 & 0 \\ 0 & 1/\alpha_2^2 & 0 \\ 0 & 0 & 1/\alpha_3^2 \end{pmatrix}$$

For an incompressible material, $\alpha_2^2 = \alpha_3^2 = 1/\alpha_1$. If the material is unstrained for $-\infty < t' < 0$, then, *relative to the reference configuration* at times $t' > 0$, $\mathbf{C}$ and $\mathbf{C}^{-1}$ are

$$\mathbf{C} = \begin{pmatrix} \alpha_1^2 & 0 & 0 \\ 0 & 1/\alpha_1 & 0 \\ 0 & 0 & 1/\alpha_1 \end{pmatrix}, \quad \mathbf{C}^{-1} = \begin{pmatrix} 1/\alpha_1^2 & 0 & 0 \\ 0 & \alpha_1 & 0 \\ 0 & 0 & \alpha_1 \end{pmatrix} \qquad (3.141)$$

For $t' \geq 0$ the material is strained, but, since this is the reference configuration, $\mathbf{C} = \mathbf{C}^{-1} = \boldsymbol{\delta}$ for $t' \geq 0$. Now consider the stress difference $\tau_{11} - \tau_{22} = T_{11} - T_{22}$. If the surfaces transverse to the direction of principal extension are unconstrained, then $T_{22} = 0$, and the total stress T_{11} is just $\tau_{11} - \tau_{22}$. From Eq. 3.139, then, it is found that

$$\tau_{11} - \tau_{22} = \int_{-\infty}^{0} \left[\mu_1\left(\alpha_1^2 - \frac{1}{\alpha_1}\right) + \mu_2\left(\frac{1}{\alpha_1^2} - \alpha_1\right)\right] dt'$$

$$= \left(\alpha_1^2 - \frac{1}{\alpha_1}\right) \int_{-\infty}^{0} \left[\mu_1 - \frac{1}{\alpha_1}\mu_2\right] dt' \qquad (3.142)$$

* In Ref. 53 the integral is written as $\mu_1'\mathbf{C}^{-1} + \mu_2'\mathbf{C}^{-1}\cdot\mathbf{C}^{-1}$. Equation 3.139 is the form used by Zapas and Craft. The two forms are equivalent, with suitable modification of the relaxation functions.

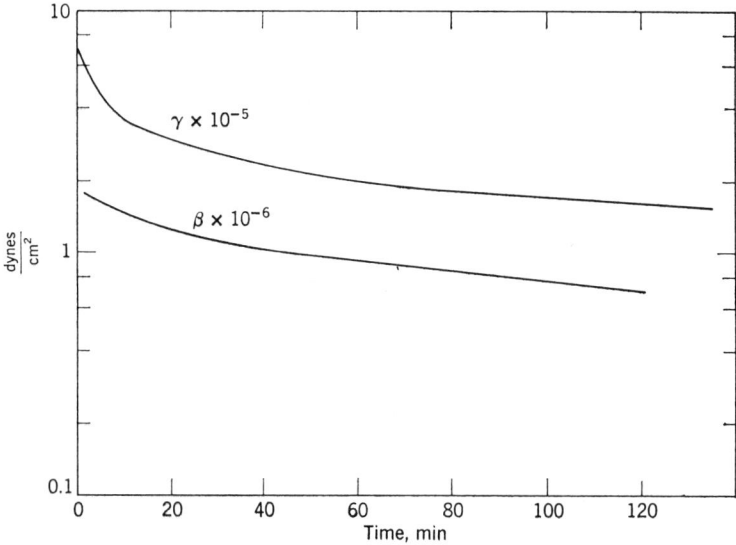

Fig. 3.7. Relaxation functions defined by Eqs. 3.142–3.144.

If two new relaxation functions are defined in such a way that

$$\mu_1 = \frac{d\gamma}{dt'} \quad \text{and} \quad \mu_2 = -\frac{d\beta}{dt'} \tag{3.143}$$

then Eq. 3.142 may be integrated to give

$$\frac{\tau_{11} - \tau_{22}}{\alpha_1^2 - \frac{1}{\alpha_1}} = \gamma(t) + \frac{1}{\alpha_1}\beta(t) \tag{3.144}*$$

The functions $\gamma(t)$ and $\beta(t)$ may be obtained by taking stress relaxation data (T_{11} vs. t) for a series of extension ratios α_1. By cross plotting the left-hand side of Eq. 3.144 against $1/\alpha_1$ at constant values of t, a series of "isochrones" are obtained from which the slope $\beta(t)$ and the intercept at $\alpha_1 = 1$, $\gamma(t) + \beta(t)$, may be obtained (assuming, of course, that the isochrones are straight lines). Figure 3.7 shows $\gamma(t)$ and $\beta(t)$ obtained from the data of Zapas and Craft.

Not all materials show agreement with Eq. 3.144, indicating usually that $\gamma(t)$ and $\beta(t)$ may also be functions of the extension ratio α_1. Zapas and Craft find, for example, that polyvinyl chloride (plasticized with 50%

---

* It is assumed that $\gamma(\infty) = \beta(\infty) = 0$. $\gamma(t)$ and $\beta(t)$ are the same functions used by Zapas and Craft.

by weight of tricresol phosphate) does not obey Eq. 3.144 without the addition of a new function, $\alpha(t)(\alpha_1^2 - 1)$, to the terms $(1/\alpha_1)\beta(t) + \gamma(t)$. This is nothing more than a reflection of the fact that $(\tau_{11} - \tau_{22})/(\alpha_1^2 - 1/\alpha_1)$ may depend upon α_1 in a manner which differs from one material to another.

Because its relaxation functions are left unspecified, the BKZ theory has much greater flexibility than other models. But the success of the BKZ theory should not be ascribed simply to its flexibility. The measure of success is not so much the ability of the model to describe accurately a given type of response, but the success achieved in using measurements of μ_1 and μ_2 from one type of deformation, and *predicting* the response to a different type of deformation, for a given material.

The success of the BKZ theory is illustrated best by the study of Zapas and Craft (55) of the deformation of undiluted polyisobutylene under conditions such that it is a rubbery elastomer. The relaxation functions are calculated from a stress relaxation experiment such as described above. These functions are then used for the prediction of stress-strain behavior in multiple-step relaxation, wherein the elastomer is stretched, relaxes for some time at constant strain, and is then given a second stretch. The agreement between theory and experiment is excellent.

Creep experiments under constant load were also performed. In this case the measured elongation was introduced into Eq. 3.139, along with the functions μ_1 and μ_2 determined from the stress relaxation at constant strain. The load was predicted and found to be in excellent agreement (within 2%) with the measured load. Creep recovery, upon removal of the load, was also predicted, and again found to be in excellent agreement with the observed recovery, even in the rapid recovery region immediately following removal of the load. Finally, a constant rate of strain experiment (actually constant speed elongation) was performed, and again the stress-strain curve predicted by the BKZ theory was in excellent agreement with the experimental data.

The dependence of the relaxation of stress on the initial elongation α_1 may be predicted, at least for small elongations, by assuming that the relaxation functions μ_1 and μ_2 depend on the strain, as well as on elapsed time. Properly stated, μ_1 and μ_2 should be functions of time and of the invariants of some strain tensor, say $\mathbf{C}^{-1}$. In the limit of infinitesimal strain, μ_1 and μ_2 should be functions only of time. This suggests that a power series expansion would be appropriate for the dependence of the relaxation functions on the invariants of $\mathbf{C}^{-1}$.

The invariants of $\mathbf{C}^{-1}$, for the extension defined by Eq. 3.141, are

$$I_{\mathbf{C}^{-1}} = 2\alpha_1 + 1/\alpha_1^2$$
$$II_{\mathbf{C}^{-1}} = 2/\alpha_1 + \alpha_1^2 \qquad (3.145)$$
$$III_{\mathbf{C}^{-1}} = \alpha_1\alpha_2\alpha_3 = 1 \quad \text{for an incompressible material}$$

CONTINUUM THEORIES

In the unstrained state, $I_{C^{-1}} = II_{C^{-1}} = 3$. An expansion for μ_1 and μ_2 might take the form

$$\mu_1(t) = \mu_{10}(t) + \mu_{12}(II_{C^{-1}} - 3)$$
$$\mu_2(t) = \mu_{20}(t)$$
(3.146)

For this special choice, it is found that three independent material functions appear, and

$$\frac{\tau_{11} - \tau_{22}}{\alpha_1^2 - 1/\alpha_1} = \gamma(t) + \frac{1}{\alpha_1}\beta(t) + (\alpha_1^2 - 1)\alpha(t)$$
(3.147)

with

$$\gamma(t) = \gamma_{10}(t) - 2\gamma_{12}(t)$$
$$\beta(t) = 2\gamma_{12}(t) + \beta_{10}(t)$$
$$\alpha(t) = \gamma_{12}(t)$$

and where

$$\frac{d\gamma_{10}}{dt} = \mu_{10}$$

$$\frac{d\gamma_{12}}{dt} = \mu_{12}$$

$$-\frac{d\beta_{10}}{dt} = \mu_{20}$$

The data of Zapas and Craft for polyvinyl chloride are described by this special choice, and the particular example simply illustrates that real behavior can be predicted by allowing the functions μ_1 and μ_2 to depend on the invariants of the deformation in a simple manner. If more terms of the series expansions for μ_1 and μ_2 are considered then, of course, more complex relaxation behavior may be predicted. DeHoff and coworkers (57) have carried out an experimental program which indicates that polyurethane, and ethylene-propylene rubber, can be described by an integral theory utilizing series expansions such as in Eq. 3.146, although more than three independent relaxation functions are required to fit their data.

It is interesting to note that Eq. 3.146 not only leads to prediction of the observed qualitative behavior of stress relaxation in strained *elastomers*, but also gives rise to non-newtonian *fluid* response to a steady simple shear deformation. Consider the deformation described by Eqs. 3.131 to 3.134. If Eq. 3.139 is taken as the constitutive relation, then the shear stress is given by

$$\tau_{12} = \int_{-\infty}^{t} (\mu_1 C_{12} + \mu_2 C_{12}^{-1})\, dt'$$
$$= \int_{-\infty}^{0} \dot\gamma t(\mu_2 - \mu_1)\, dt' + \int_{0}^{t} \dot\gamma s(\mu_1 - \mu_2)\, dt'$$
(3.148)

If $s = -(t - t')$ is introduced for the integration variable, the stress takes the form

$$\tau_{12} = \dot\gamma t \int_{-\infty}^{-t} m(s)\, ds - \dot\gamma \int_{-t}^{0} s m(s)\, ds \qquad (3.149)$$

where $m(s) = \mu_2 - \mu_1$. If Eq. 3.146 is used, then

$$m(s) = m_1(s) + m_2(s)(II_{C^{-1}} - 3) \qquad (3.150)$$

where $m_1(s)$ and $m_2(s)$ may be identified in terms of μ_{10}, μ_{20}, and μ_{12}, and $II_{C^{-1}}$ is given by

$$\begin{aligned} II_{C^{-1}} &= 3 + s^2 \dot\gamma^2 \qquad \text{for} \quad t' > 0 \\ II_{C^{-1}} &= 3 + t^2 \dot\gamma^2 \qquad \text{for} \quad t' \leqslant 0 \end{aligned} \qquad (3.151)$$

The stress now becomes

$$\begin{aligned} \tau_{12} = {} & \dot\gamma t \int_{-\infty}^{-t} [m_1(s) + t^2 \dot\gamma^2 m_2(s)]\, ds \\ & - \dot\gamma \int_{-t}^{0} [s m_1(s) + s^3 \dot\gamma^2 m_2(s)]\, ds \end{aligned} \qquad (3.152)$$

The viscosity will be given by $\tau_{12}/\dot\gamma$ when equilibrium is achieved at $t = \infty$. Hence the viscosity may be written

$$\eta = \int_{0}^{-\infty} [s m_1(s) + s^3 \dot\gamma^2 m_2(s)]\, ds \qquad (3.153)$$

At vanishingly small shear rates, the zero shear viscosity is seen to be*

$$\eta_0 = \int_{0}^{-\infty} s m_1(s)\, ds \qquad (3.154)$$

At higher shear rates, the appearance of $\dot\gamma$ in the integrand of Eq. 3.153 gives a non-newtonian viscosity, and η takes the form†

$$\eta = \eta_0 [1 - (\lambda \dot\gamma)^2] \qquad (3.155)$$

where a "timelike" parameter λ appears, defined by

$$\lambda^2 = -\int_{0}^{-\infty} s^3 m_2(s)\, ds / \eta_0 \qquad (3.156)$$

If Eqs. 3.150 and 3.146 are compared it may be seen that $-m_2(s) = \mu_{12}(s)$. It is clear, then, that non-newtonian shear behavior, in the BKZ elastic fluid, is associated with strain invariant-dependent relaxation functions.

* Eq. 3.154 is identical with Eq. 3.53, if s is replaced by $-s$, and $\mu = m_1$.
† Since Eq. 3.150 can only be a good approximation for nearly strain-free conditions, it is not fair to criticize Eq. 3.155 for a negative viscosity at large shear rates.

By contrast the Bird-integral model defined in Eq. 3.130 shows non-newtonian behavior because of a strain *rate* invariant-dependent relaxation function.

Earlier in this section the relationship of integral theories to classical *viscoelasticity*, in particular, the Boltzmann superposition integral, was described. It is (hopefully) instructive to point out a relationship of the BKZ theory to classical *elasticity*, that is, to the theory of perfectly elastic materials in which no flow occurs, and in which relaxation is instantaneous. The discussion will be simplified by considering homogeneous isotropic incompressible materials, and by performing the development in terms of cartesian components. More general discussions are available elsewhere (3, 4).

As a starting point one defines a perfectly elastic material as one in which deformation causes only reversible changes in internal energy. Hence the internal energy is a conserved quantity, and one may define an "elastic potential" or "strain energy function" U, such that the rate at which work is done on the material by the stresses is related to the potential function by

$$\frac{DU}{Dt} = \tau_{kl} \frac{\partial v_l}{\partial x_k} \tag{3.157}$$

In the simple material to be considered here, U is assumed to be a function only of some strain measure. For example, the displacement gradient $F_{kk} = \partial \xi_k / \partial x_k$ may be used, where ξ and $\mathbf{x}$ are the places occupied by material particles at the times t' and t, respectively. (See Appendix B for a more detailed discussion of strain.) Then

$$\frac{DU}{Dt} = \frac{\partial U}{\partial F_{kk}} \frac{DF_{kk}}{Dt} \tag{3.158}$$

Consider, for a moment, the quantity $\mathbf{F}^{-1}\mathbf{F} = \boldsymbol{\delta}$, or

$$\frac{\partial x_l}{\partial \xi_i} \frac{\partial \xi_i}{\partial x_k} = \delta_{lk} \tag{3.159}$$

Since $\boldsymbol{\delta}$ is a constant its material time derivative vanishes, and, using the fact that $\partial \xi_i / \partial x_k = F_{ik}$, it is seen that

$$\frac{\partial x_l}{\partial \xi} \frac{DF_{ik}}{Dt} = -F_{ik} \frac{D}{Dt} \frac{\partial x_l}{\partial \xi_i} \tag{3.160}$$

By definition of the material derivative it may be shown that

$$\frac{D}{Dt} \frac{\partial x_l}{\partial \xi_i} = \frac{\partial}{\partial \xi_i} \frac{Dx_l}{Dt} = \frac{\partial v_l}{\partial \xi_i} = \frac{\partial v_l}{\partial x_k} \frac{\partial x_k}{\partial \xi_i} \tag{3.161}$$

If this is introduced into Eq. 3.160 and note is taken of the fact that

$F_{ik}\, \partial x_k/\partial \xi_i = 1$, it may be seen that

$$\frac{\partial x_l}{\partial \xi_i}\frac{DF_{ik}}{Dt} = -F_{ik}\frac{\partial x_k}{\partial \xi_i}\frac{\partial v_l}{\partial x_k} = -\frac{\partial v_l}{\partial x_k} \qquad (3.162)$$

If this is multiplied through by $\partial \xi_k/\partial x_l$ it follows that

$$\frac{\partial \xi_k}{\partial x_l}\frac{\partial x_l}{\partial \xi_i}\frac{DF_{ik}}{Dt} = -\frac{\partial \xi_k}{\partial x_l}\frac{\partial v_l}{\partial x_k} = -F_{kl}\frac{\partial v_l}{\partial x_k}$$

or

$$\frac{DF_{kk}}{Dt} = -F_{kl}\frac{\partial v_l}{\partial x_k} \qquad (3.163)$$

(The relation $\dfrac{\partial \xi_k}{\partial x_l}\dfrac{\partial x_l}{\partial \xi_i} = \dfrac{\partial \xi_k}{\partial \xi_i} = \delta_{ki}$ has been used here.)

Equation 3.157 may now be written as

$$\tau_{kl}\frac{\partial v_l}{\partial x_k} = \frac{\partial U}{\partial F_{kk}}\frac{DF_{kk}}{Dt} = -\frac{\partial U}{\partial F_{kk}}F_{kl}\frac{\partial v_l}{\partial x_k}$$

or simply

$$\tau_{kl} = -F_{kl}\frac{\partial U}{\partial F_{kk}} \qquad (3.164)$$

In place of the components of the displacement gradient tensor **F**, the components of the strain tensor **C** may be introduced as the strain measures upon which U depends, and the stress becomes

$$\tau_{kl} = -F_{kl}\frac{\partial U}{\partial C_{mm}}\frac{\partial C_{mn}}{\partial F_{kk}} \qquad (3.165)$$

Using Eq. 3.33 (with a notation change), it may be seen that

$$\begin{aligned}F_{kl}\frac{\partial C_{mn}}{\partial F_{kk}} &= F_{kl}\frac{\partial}{\partial F_{kk}}(F_{im}F_{jn}\delta_{ij})\\ &= F_{kl}F_{im}\delta_{nk}\delta_{ik} + F_{kl}F_{jn}\delta_{jk}\delta_{mk}\\ &= C_{lm}\delta_{nk} + C_{ln}\delta_{mk}\end{aligned} \qquad (3.166)$$

Hence

$$\tau_{kl} = -\frac{\partial U}{\partial C_{mn}}(C_{lm}\delta_{nk} + C_{ln}\delta_{mk})$$

$$= -C_{lm}\frac{\partial U}{\partial C_{km}} - C_{ln}\frac{\partial U}{\partial C_{kn}}$$

and if the dummy index n in the second term on the right-hand side is changed to m

$$\tau_{kl} = -2C_{lm}\frac{\partial U}{\partial C_{km}} \qquad (3.167)$$

CONTINUUM THEORIES

While the path to Eq. 3.167 is algebraically tortuous, it has led to the desired (although still intermediate) result of a constitutive equation for τ in terms of the strain tensor $\mathbf{C}$ and in terms of the strain energy function U. The remaining steps are toward putting this equation into a form more suitable to interpretation and more amenable to computation.

Since U is a scalar quantity it can depend on $\mathbf{C}$ only through the invariants of $\mathbf{C}$:

$$I_\mathbf{C} = C_{km}\delta_{km}$$
$$II_\mathbf{C} = \tfrac{1}{2}(C_{kk}C_{mm} - C_{km}C_{km}) \qquad (3.168)$$
$$III_\mathbf{C} = 1 \quad \text{for an incompressible material}$$

The derivative of U is now written as

$$\frac{\partial U}{\partial C_{km}} = \frac{\partial U}{\partial I}\frac{\partial I}{\partial C_{km}} + \frac{\partial U}{\partial II}\frac{\partial II}{\partial C_{km}} \qquad (3.169)$$

(The subscript $\mathbf{C}$ has been dropped from I and II for convenience.) From Eq. 3.168 it can be seen that

$$\frac{\partial I}{\partial C_{km}} = \delta_{km}$$
$$\frac{\partial II}{\partial C_{km}} = I\delta_{km} - C_{km} \qquad (3.170)$$

Upon introducing these results into Eq. 3.167, one obtains a constitutive equation of the form

$$\tau_{kl} = -2C_{lm}\delta_{km}\frac{\partial U}{\partial I} - 2C_{lm}\frac{\partial U}{\partial II}(I\delta_{km} - C_{km})$$

or

$$\tau_{kl} = -2\left[\frac{\partial U}{\partial I} + I\frac{\partial U}{\partial II}\right]C_{kl} - 2\frac{\partial U}{\partial II}C_{km}C_{lm} \qquad (3.171)$$

For the sake of showing a correspondence to the BKZ theory one further algebraic step must be suffered. The Cayley-Hamilton theorem (Appendix A, Eq. A.45a) may be written as

$$\mathbf{C}^3 - I\mathbf{C}^2 + II\mathbf{C} - III\boldsymbol{\delta} = 0$$

If this is multiplied by $\mathbf{C}^{-1}$ the result is

$$\mathbf{C}^2 - I\mathbf{C} + II\boldsymbol{\delta} - III\mathbf{C}^{-1} = 0$$

or, in cartesian components

$$C_{km}C_{lm} - C_{kl}I + \delta_{kl}II - IIIC_{kl}^{-1} = 0$$

Hence the second-order term in Eq. 3.171 may be eliminated to give

(upon setting $III = 1$)

$$\tau_{kl} = -2\frac{\partial U}{\partial I} C_{kl} + 2\frac{\partial U}{\partial II} C_{kl}^{-1} - 2II \frac{\partial U}{\partial II}\delta_{kl} \qquad (3.172)$$

The term $2II\delta_{kl}\,\partial U/\partial II$ is an isotropic component of stress. Since the material is assumed to be incompressible, its deformation behavior is unaffected by isotropic stress components, and this term may be included as part of the pressure in the total stress $\mathbf{T} = -p\boldsymbol{\delta} + \boldsymbol{\tau}$. The final result, then, is given in the form

$$\boldsymbol{\tau} = -2\frac{\partial U}{\partial I}\mathbf{C} + 2\frac{\partial U}{\partial II}\mathbf{C}^{-1} \qquad (3.173)$$

If the invariants of $\mathbf{C}^{-1}$ are introduced, using (as can be easily proven) $I_\mathbf{C} = II_{\mathbf{C}^{-1}}$ and $II_\mathbf{C} = I_{\mathbf{C}^{-1}}$, then the constitutive equation becomes

$$\boldsymbol{\tau} = -2\left(\frac{\partial U}{\partial II_{\mathbf{C}^{-1}}}\mathbf{C} - \frac{\partial U}{\partial I_{\mathbf{C}^{-1}}}\mathbf{C}^{-1}\right) \qquad (3.174)$$

which is the form due to Finger (58).

The strain energy function $U(I, II)$ may be taken to depend upon the invariants of $\mathbf{C}^{-1}$. If $U(I, II)$ is specified, then the material coefficients $\partial U/\partial I$ and $\partial U/\partial II$ are obtained, and it is clear that these coefficients are not independent, being derived from a single function U.

While the superficial relationship of Eq. 3.174 to Eq. 3.139 and 3.140 is apparent, there is no guarantee that a function analogous to U must exist is a *dissipative* material, i.e., in a *fluid*. In the BKZ theory, which is an elastic fluid theory, U is a relaxing time-dependent potential function. Hence $U(t - t', I, II)$ is a non-equilibrium property which depends on the state of the material with respect to equilibrium, given through the dependence on I and II, and also on the elapsed time $t - t'$. This latter feature gives U the aspect of a memory function. Bernstein, Kearsley, and Zapas (59) discuss the thermodynamic aspects of the development of an elastic fluid theory, and present a non-equilibrium thermodynamic foundation for the theory.

The ultimate test of the existence of an elastic fluid potential function is the demonstration that some function U, when introduced into Eqs. 3.139 and 3.140, allows the prediction of mechanical response in a real material. Zapas (60) has carried out an experimental program which suggests a potential function of the form

$$U = \frac{\alpha'}{2}(II_{\mathbf{C}^{-1}} - 3)^2 + 4.5\beta' \ln\left(\frac{I_{\mathbf{C}^{-1}} + II_{\mathbf{C}^{-1}} + 3}{9}\right)$$

$$+ 24(\beta' - c')\ln\frac{II_{\mathbf{C}^{-1}} + 15}{I_{\mathbf{C}^{-1}} + 15} + c'(II_{\mathbf{C}^{-1}} - 3) \qquad (3.175)$$

Creep and relaxation data for vulcanized butyl rubber are described by the BKZ theory, with this choice for U.

In a simple shear flow η and ψ_{12} are predicted to have the forms

$$\eta = 2 \int_0^\infty \left[\alpha'(s)\dot{\gamma}^2 s^2 + \frac{9\beta'(s)}{9 + 2\dot{\gamma}^2 s^2} + c'(s) \right] s\, ds \tag{3.176}$$

$$\psi_{12} = -2 \int_0^\infty \left[\alpha'(s)\dot{\gamma}^2 s^2 + \frac{9\beta'(s)}{9 + 2\dot{\gamma}^2 s^2} + c'(s) \right] s^2\, ds \tag{3.177}$$

Unless $\alpha'(s)$ is zero for all s, both material functions will increase with $\dot{\gamma}$ at very large values of $\dot{\gamma}$, contrary to experimental evidence. Hence this choice for U cannot predict the behavior of real polymer *fluids* unless α' vanishes.

For very high shear rates the viscosity is given, in the limit of infinite $\dot{\gamma}$, and assuming $\alpha' = 0$, by

$$\lim_{\dot{\gamma} \to \infty} \eta = 2 \int_0^\infty c'(s) s\, ds = \eta_\infty \tag{3.178}$$

where η_∞ is the "upper newtonian viscosity." At the other extreme the zero shear viscosity is found to be

$$\lim_{\dot{\gamma} \to 0} \eta = 2 \int_0^\infty \beta'(s) + c'(s)\, s\, ds \tag{3.179}$$

Except in quite dilute solutions η_∞ is orders of magnitude smaller than η_0 (29). Hence, in most polymer solutions, and certainly in polymer melts, the viscosity must be dominated by the relaxation function $\beta'(s)$. To a good approximation, then, the potential function suggested by Zapas becomes, for polymer fluids

$$U = 4.5\beta' \ln\left(\frac{I + II + 3}{9}\right) + 24\beta' \ln\left(\frac{II + 15}{I + 15}\right) \tag{3.180}$$

and η and ψ_{12} are reduced to

$$\eta = 2 \int_0^\infty \frac{\beta'(s)}{1 + \tfrac{2}{9}\dot{\gamma}^2 s^2}\, s\, ds \tag{3.181}$$

and

$$\psi_{12} = 2 \int_0^\infty \frac{\beta'(s)}{1 + \tfrac{2}{9}\dot{\gamma}^2 s^2}\, s^2\, ds \tag{3.182}$$

Zapas further supports his choice for U with data for a fairly viscous polyisobutylene solution. The complex viscosity was measured over a range of frequencies and the material functions in U were fitted to the data.

With these functions the steady shear viscosity was predicted and found to be in excellent agreement with data obtained in the range $\eta/\eta_0 > 0.1$.

More complete experimental studies will be required before it is possible to argue in favor of the BKZ theory, with U as a single potential from which μ_1 and μ_2 may be derived, as a constitutive equation applicable to polymer solutions and melts over a wide range of shear rates. It does, however, seem to be quite successful in describing both the behavior of rubbery elastomers in creep and elongation experiments, and the non-newtonian viscosity of polymer solutions.

REFERENCES

1. Truesdell, C. A., *Proceedings of the Fourth International Congress on Rheology*, Pt. 2, E. H. Lee, Ed., Interscience, New York, 1965, p. 3.
2. Oldroyd, J. G., *Proc. Roy. Soc. (London)*, **A200**, 523 (1950).
3. Eringen, A. C., *Nonlinear Theory of Continuous Media*, McGraw-Hill, New York, 1962, p. 93.
4. Prager, W., *Introduction to Mechanics of Continua*, Ginn, Boston, 1961, pp. 154–156.
5. Noll, W., *Arch. Rat. Mech. Anal.* **2**, 197 (1958).
6. Coleman, B. D., and W. Noll, *Arch. Rat. Mech. Anal.*, **3**, 289 (1959).
7. Rivlin, R. S., and J. L. Ericksen, *J. Rat. Mech. Anal.*, **4**, 323 (1955).
8. Coleman, B. D., and W. Noll, *Rev. Mod. Phys.*, **33**, 239 (1961).
9. Fredrickson, A. G., *Chem. Eng. Sci.*, **17**, 155 (1962).
10. Bogue, D. C., *Ind. Eng. Chem. Fundamentals*, **5**, 253 (1966).
11. Bogue, D. C., and J. O. Doughty, *Ind. Eng. Chem. Fundamentals*, **5**, 243 (1966).
12. Slattery, J. C., and R. B. Bird, *Chem. Eng. Sci.*, **16**, 231 (1961).
13. Sutterby, J. L., *Trans. Soc. Rheol.*, **9:2**, 227 (1965).
14. Dodge, D. W., and A. B. Metzner, *A.I.Ch.E. J.*, **5**, 189 (1959); see also Metzner, A. B., and J. C. Reed, *A.I.Ch.E. J.*, **1**, 434 (1955).
15. Metzner, A. B., and R. Otto, *A.I.Ch.E. J.*, **3**, 3 (1957).
16. Shah, M. J., E. E. Petersen, and A. Acrivos, *A.I.Ch.E. J.*, **8**, 542 (1962).
17. Christopher, R., and S. Middleman, *Ind. Eng. Chem. Fundamentals*, **4**, 422 (1965).
18. Griffith, R. M., *Ind. Eng. Chem. Fundamentals*, **1**, 180 (1962).
19. Bird, R. B., *Can. J. Chem. Eng.*, **44**, 161 (1965).
20. Barnett, S., A. Humphrey, and M. Litt, *A.I.Ch.E. J.*, **12**, 253 (1966).
21. McEachern, D. W., *A.I.Ch.E. J.*, **12**, 328 (1966).
22. Sadowski, T. J., and R. B. Bird, *Trans. Soc. Rheol.*, **9:2**, 243 (1965).
23. Sadowski, T. J., *Trans. Soc. Rheol.*, **9:2**, 251 (1965).
24. Prandtl, L., *Physik. Bl.*, **5**, 161 (1949).
25. Ree, F. H., T. Ree, and H. Eyring, *Ind. Eng. Chem.*, **50**, 1036 (1958).
26. Ree, T., and H. Eyring, *J. Appl. Phys.*, **26**, 793, 800 (1955).
27. Powell, R. E., and H. Eyring, *Nature*, **154**, 427 (1944).
28. Merrill, E. W., H. S. Mickley, A. Ram, and W. H. Stockmayer, *Trans. Soc. Rheol.*, **6**, 119 (1962); also *J. Polymer Sci. A*, **1**, 1201 (1963).
29. Brodnyan, J. G., F. Gaskins, and W. Philippoff, *Trans. Soc. Rheol.*, **1**, 109 (1957).
30. Sutterby, J. L., *A.I.Ch.E. J.*, **12**, 63 (1966).

31. Meter, D. M., and R. B. Bird, *A.I.Ch.E. J.*, **10**, 878 (1964).
32. Gillespie, T., *J. Colloid Interface Sci.*, **22**, 133 (1966).
33. Spencer, R. S., and R. Dillon, *J. Colloid Sci.*, **4**, 241 (1949).
34. Seely, G. R., *A.I.Ch.E. J.*, **10**, 56 (1964).
35. Maxwell, J. C., *Phil. Trans. Roy. Soc. (London)*, **157**, 49 (1867).
36. Spriggs, T. W., *Chem. Eng. Sci.*, **20**, 931 (1965).
37. Oldroyd, J. G., *Quart. J. Mech. Appl. Math.*, **4**, 271 (1951).
38. Oldroyd, J. G., *Proc. Roy. Soc. (London)*, **A245**, 278 (1958).
39. Rouse, P. E., *J. Chem. Phys.*, **21**, 1272 (1953).
40. Zimm, B. H., *J. Chem. Phys.*, **24**, 269 (1956).
41. Onogi, S., T. Fujii, H. Kato, and S. Ogihara, *J. Phys. Chem.*, **68**, 1598 (1964).
42. DeWitt, T., H. Markovitz, F. Padden, and L. Zapas, *J. Colloid. Sci.*, **10**, 174 (1955).
43. Tobolsky, A. V., *Properties and Structure of Polymers*, Wiley, New York, 1960, p. 15.
44. Tobolsky, A. V., and K. Murakami, *J. Polymer Sci.*, **40**, 443 (1959).
45. Abramowitz, M., Ed., *Handbook of Mathematical Functions*, NBS Appl. Math. Series 55, U.S. Government Printing Office, Washington, D.C., 1964, p. 1029.
46. Alfrey, T., *Mechanical Behavior of High Polymers*, Interscience, New York, 1948, p. 533.
47. Ferry, J. D., and M. L. Williams, *J. Colloid Sci.*, **7**, 347 (1952).
48. Abramowitz, M., see ref. 45, p. 253.
49. Tobolsky, A. V., B. Dunell, and R. Andrews, *Textile Res. J.*, **21**, 404 (1951).
50. Tobolsky, A. V., *J. Am. Chem. Soc.*, **74**, 3786 (1952).
51. Catsiff, E., and A. V. Tobolsky, *J. Colloid Sci.*, **10**, 375 (1955).
52. Boltzmann, L., *Pogg. Ann. Phys.*, **7**, 624 (1876).
53. Spriggs, T. W., J. D. Huppler, and R. B. Bird, *Trans. Soc. Rheol.*, **10:1**, 191 (1966).
54. Huppler, J. D., Paper presented to Society of Rheology, Atlantic City, Oct., 1966.
55. Zapas, L., and T. Craft, *J. Res. Natl. Bur. Std.*, **69A**, 541 (1965).
56. Bernstein, B., E. Kearsley, and L. Zapas, *Trans. Soc. Rheol.*, **7**, 391 (1963).
57. DeHoff, P. H., G. Lianis, and W. Goldberg, *Trans. Soc. Rheol.*, **10:1**, 385 (1966).
58. Finger, J., *Akad. Wiss. Wein. Sitzber.*, **103**, 1073 (1894); also ref. 4, p. 209 (but note a sign difference in the definition of II).
59. Bernstein, B., E. A. Kearsley, and L. Zapas, *J. Res. Natl. Bur. Std.*, **68B**, 103 (1964).
60. Zapas, L., *J. Res. Natl. Bur. Std.*, **70A**, 525 (1966).

4

Molecular Theories of Polymer Rheology

Up to this point the discussion of rheological principles and methods has not required any reference to structure in a material. But clearly a material, even a homogeneous material, does have a structure endowed to it by its constituent molecules, and the rheological response of the material is related to its molecular structure. Whether a material is molecular or continuous is a semantic matter until one attempts to describe the response of the material through the use of theoretical ideas which depend specifically upon the physics of molecules, or the physics of continua.

In recent years a number of successful developments have been made in the treatment of rheological response based upon mathematical models of the structure and dynamics of long chain molecules. This chapter will present a review of the fundamental principles and ideas upon which these developments have been based. Some specific examples of the success of molecular theories will be given. In Chapter 5 more extensive examples will be used to show the dependence of rheological properties on molecular characteristics.

I. THE POLYMER MOLECULE

The word polymer is generally used to describe a large molecule made up of many chemically linked repeating units of smaller molecules. These smaller units, called "monomers," can combine through a variety of chemical reactions which are usually peculiar to the type of monomer involved. For example, "addition polymerization" can occur among monomer units containing a carbon–carbon double bond to lead to polymers of the type

$$H\text{—}[R]_n\text{—}H$$

where [R] represents some repeating unit formed from the monomer, and [ ]$_n$ means that n monomers are linked in a linear chain. An example of such a polymerization is that of styrene monomer to polystyrene:

$$\underset{\substack{|\\H}}{\overset{\substack{H\\|}}{C}}=\underset{\substack{|\\\phi}}{\overset{\substack{H\\|}}{C}} + \underset{\substack{|\\H}}{\overset{\substack{H\\|}}{C}}=\underset{\substack{|\\\phi}}{\overset{\substack{H\\|}}{C}} \longrightarrow -\underset{\substack{|\\H}}{\overset{\substack{H\\|}}{C}}-\underset{\substack{|\\\phi}}{\overset{\substack{H\\|}}{C}}-\underset{\substack{|\\H}}{\overset{\substack{H\\|}}{C}}-\underset{\substack{|\\\phi}}{\overset{\substack{H\\|}}{C}}-$$

The dimer thus formed may then react on both ends with more styrene to produce very long molecules. The "degree of polymerization," which is just the number of monomers in the chain, may reach ten thousand before the reaction terminates, and so molecular weights may be of the order of millions.

The peculiarities of the polymerization reaction may sometimes lead to the removal of a hydrogen atom internal to the chain, with the result that a "branch chain" may begin and grow to some appreciable length. As branching alters mechanical properties the existence of significant degrees of branching may or may not be desirable, and one attempts to control branching through the use of specific types of catalysts and through careful control of reaction temperature and pressure. If branches of two or more chains should join the polymer may become "crosslinked" and take on very different properties from the linear polymer, due to the formation of two- or three-dimensional structure in the bulk material.

The degree of polymerization depends upon factors which are often difficult to control in commercial reactors. Thus most commercial polymers are mixtures of chains of various molecular weights, and one may define a molecular weight distribution function and various molecular weight averages for a polymer sample. Suppose, for example, that one denotes the *number* fraction of chains of molecular weight in the range M to $M + dM$ by $f(M)$. Then the "number average molecular weight" M_n is defined by

$$M_n = \int_0^\infty Mf(M)\,dM \qquad (4.1)*$$

and the "weight average molecular weight" M_w is defined by

$$M_w = \int_0^\infty M^2 f(M)\,dM \Big/ \int_0^\infty Mf(M)\,dM \qquad (4.2)$$

Other averages may be defined, and the relationships among the various averages depend upon the particular molecular weight distribution.

An extensive discussion of methods of measuring various molecular weight averages and the molecular weight distribution, and a treatment

* Although molecular weight is a *discrete* variable, a *continuous* distribution function is used. The error is negligible in view of the high molecular weights of interest.

of the effect of various polymerization kinetics on the molecular weight distribution, are given by Flory (1). Such detail will not be required here. It should be pointed out, however, that one reason for defining different molecular weight averages lies in the fact that certain experimental techniques are sensitive to different moments of the molecular weight distribution, that is, to different average molecular weights. For this reason one must be careful in stating "a" molecular weight to make clear what kind of molecular weight has been determined.

Billmeyer (2) describes very concisely some methods of determining molecular weight averages. One such method depends upon a measurement of osmotic pressure of polymer solutions. Since this is a colligative property, it is sensitive to the *number* of polymer molecules in solution and to the number of monomer units present. Thus osmotic pressure gives the number average molecular weight M_n. On the other hand, light scattering from polymer solutions, which is the basis of another common technique of molecular weight determination, depends upon the *mass* of the scattering particle, and the resulting molecular weight determination actually gives the weight average M_w.

For a monodisperse polymer, of course, there is only one molecular weight to be measured and all techniques should give the same value, except for experimental errors. For polydisperse polymers the weight average M_w is found to be larger than the number average M_n. The ratio M_w/M_n is a measure, although not a unique one, of the breadth of the molecular weight distribution. Values of M_w/M_n in the range 10 to 30 are not uncommon, for example, in commercial polyethylene (3). Well-controlled polymerizations can yield polymers with $M_w/M_n < 1.1$, as for example, in polystyrene (4).

A. Spatial Configuration of a Polymer Chain

While the chemical *formula* for a linear polymer might be written as

$$H—R—R—\cdots—R—R—H$$

it is important that one not confuse this with the *configuration* of the polymer, that is, the spatial distribution of the monomeric elements. In general the monomer elements must be related spatially to one another in such a way as to preserve certain bond angle restrictions, yet within these restrictions a large number of spatial configurations is possible. Figure 4.1 shows the locus of positions of a simple three-carbon chain. One can imagine that a thousand-carbon chain must have an enormous number of possible configurations. Furthermore, thermal fluctuations (Brownian motion) must cause such a molecule to continually assume different possible configurations. For these reasons the spatial configura-

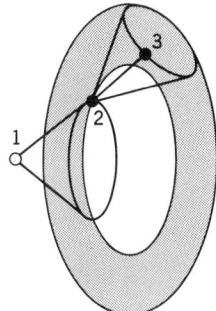

Fig. 4.1. The shaded area gives the locus of possible configurations of atom number 3.

tion of a polymer molecule has meaning only in a statistical sense, and certain statistical measures are used to describe the "shape" of a polymer molecule.

In the absence of bias due to non-isotropic stress or electrostatic fields there is no reason for a long polymer molecule to adopt a statistical configuration other than spherical. One can define a radius for this sphere and use this as a measure of the size of the polymer molecule. In the statistical sense of this model it is possible to demonstrate that the radius of a polymer molecule is proportional to the square root of the degree of polymerization, or (1, p. 408)

$$R^2 = \text{(constant)} \, N \qquad (4.3)$$

The derivation of this equation is based upon a purely geometrical argument which ignores the volume occupied by the atoms of the polymer. This volume excludes certain configurations and causes R^2 to increase somewhat more rapidly than linearly with N. This is called an "excluded volume" effect.

In a real polymer molecule there is also an effect due to inter- and intra-molecular forces exerted upon a monomer unit by its neighbors. Because these forces might differ in a given polymer immersed in different solvents, the molecular radius is not a unique property of the polymer, but depends also on the polymer's environment.

One of the primary statistical measures of the spatial configuration of a polymer molecule is the "chain-end distribution function," $P(x, y, z)$. This may be defined as the probability that the ends of a linear polymeric chain are separated by a vector $\mathbf{r}$ with components (x, y, z) in a coordinate system arbitrarily placed at one end of the chain. The magnitude of $\mathbf{r}$ may be called the end-to-end length of the molecule, and it is sometimes used as a measure of the radius R written in Eq. 4.3.

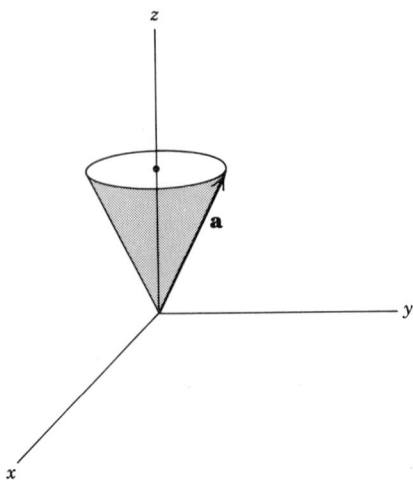

Fig. 4.2. The vector **a** represents one bond of a freely oriented chain.

A common model for the polymer is that of a freely oriented chain, a chain in which free rotation may occur about bonds. The end of such a chain is illustrated in Fig. 4.2. The bond between monomers will be denoted by the vector **a**. With respect to an arbitrary coordinate system, such as the one drawn in Fig. 4.2, it can be seen that the symmetry resulting from free rotation leads to a zero mean value of a_x and a_y. The root mean square values will not vanish, however, but are given by (1, p. 403)

$$\langle a_x \rangle^2 = \langle a_y \rangle^2 = \langle a_z \rangle^2 = a^2/3 \tag{4.4}$$

The chain-end distribution function for a polymer with N monomer units may be denoted $P(\mathbf{r}, N)$. The vector end-to-end distance for some configuration **r** is just the sum of bond vectors

$$\mathbf{r} = \sum_{i=1}^{N} \mathbf{a}_i$$

and so the addition of one more monomer has the effect of adding a vector **a** to **r**. This can be stated, for some particular configuration, as

$$P(\mathbf{r}, N+1) = P(\mathbf{r} + \mathbf{a}, N) \tag{4.5}$$

If the function $P(\mathbf{r} + \mathbf{a}, N)$ is now expanded in terms of $P(\mathbf{r}, N)$ the result is

$$P(\mathbf{r} + \mathbf{a}, N) = P(\mathbf{r}, N) + \left(\frac{\partial P}{\partial x} a_x + \frac{\partial P}{\partial y} a_y + \frac{\partial P}{\partial z} a_z \right)$$

$$+ \frac{1}{2} \left(\frac{\partial^2 P}{\partial x^2} a_x^2 + \frac{\partial^2 P}{\partial y^2} a_y^2 + \frac{\partial^2 P}{\partial z^2} a_z^2 + 2 \frac{\partial^2 P}{\partial x \partial y} a_x a_y + \cdots \right)$$

$$+ \text{ higher order terms in } \mathbf{a} \tag{4.6}$$

If this is now averaged over all possible configurations it follows that

$$\langle P(\mathbf{r} + \mathbf{a}, N)\rangle = P(r, N + 1) = P(r, N) + \frac{a^2}{6} \nabla^2 P \qquad (4.7)$$

where use has been made of Eq. 4.4 and the vanishing of $\langle \mathbf{a}\rangle$. If a^2 is much smaller than r^2, as would be true for large N, then the higher order terms may be neglected with small effect. The notation $P(r, N)$ is now used to denote the configurational average, $\langle P(\mathbf{r}, N)\rangle$.

In the limit of very large N the difference $P(r, N + 1) - P(r, N)$ will approach the derivative $\partial P/\partial N$, so that Eq. 4.7 may be written as

$$\frac{\partial P}{\partial N} = \frac{a^2}{6} \nabla^2 P \qquad (4.8)$$

the solution of which is

$$P(r) = (\tfrac{3}{2}\pi N a^2)^{3/2} \exp(-3r^2/2Na^2) \qquad (4.9)$$

This is a Gaussian distribution function, and it is a spherically symmetric function of $r^2 = x^2 + y^2 + z^2$. It is in this statistical sense that one speaks of a very large molecule as having a spherical configuration. If the "radius" R is now taken to be $\langle r^2\rangle^{1/2}$, it follows from

$$\langle r^2\rangle = \int_0^\infty r^2 P(r) 4\pi r^2 \, dr \qquad (4.10)$$

that

$$R = \sqrt{N}\, a \qquad (4.11)$$

which is of the form of Eq. 4.3.

If the chain is subject to bond angle restrictions, and so is not freely oriented, the calculations above do not hold. It may be shown (1, p. 413), however, that as long as the molecule is long the chain-end distribution is Gaussian to good approximation, and Eq. 4.3 still holds.

B. Elasticity of a Polymer Chain

Suppose one of the bonds of a long chain is acted upon by a force $-\mathbf{F}$, and suppose further that an x-axis is oriented colinear with $\mathbf{F}$. With reference to Fig. 4.3 it can be seen that, in general, the effect of such a force is to displace a monomer unit by a distance a_x in the direction of the force. The work required to support this displacement is simply $a_x F$, and if $a_x = 0$ is the equilibrium configuration, then the potential energy associated with this displacement is also $a_x F$. If one assumes that a Boltzmann type of energy distribution governs departure from equilibrium, then the probability that a displacement in the range a_x to $a_x + da_x$

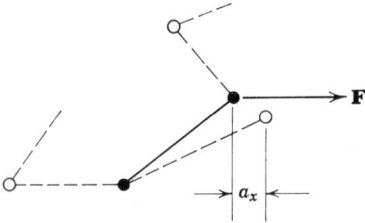

Fig. 4.3. Displacement of a monomer unit by a force **F**.

has occurred is proportional to $\exp(a_x F/kT)\, da_x$. The average displacement of this particular bond must then be

$$\bar{a}_x = \int_{-a}^{a} a_x \exp(a_x F/kT)\, da_x \Big/ \int_{-a}^{a} \exp(a_x F/kT)\, da_x \quad (4.12)$$

$$= a[\coth(aF/kT) - kT/aF]$$

If the potential energy is small in comparison with the energy due to thermal fluctuations, a power series approximation of the bracketed term may be used with the result

$$\bar{a}_x = a^2 F/3kT \quad (4.13)$$

If each bond is now assumed to behave in a similar manner, on the average, then the end-to-end component of the entire chain will be increased by an amount $\langle a_x \rangle$, given by

$$\langle a_x \rangle = N\bar{a}_x = a^2 FN/3kT \quad (4.14)$$

If F is solved for one finds

$$F = 3kT\langle a_x \rangle/Na^2 \quad (4.15)$$

The approximation used in expanding the function in Eq. 4.12 is valid if $aF/kT \ll 1$. Equation 4.15 indicates that this will be so if $3\langle a_x \rangle/Na \ll 1$. Since Eq. 4.4 gives $\langle a_x \rangle$ as of the same order of magnitude as a, it can be seen that these calculations will hold so long as $N \gg 1$, which, of course, is always true for a high polymer.

Equation 4.15 is analogous to Hooke's law for an elastic spring. It states that the end-to-end separation is maintained by a force proportional to the separation. The spring constant, $3kT/Na^2$, is not associated with any *mechanical* property of a polymer chain, however. It is a *statistical* property, in the sense that the "springiness" is associated with departure from an equilibrium configuration which is statistically more probable than other possible chain geometries. This formal relationship of the deterministic mechanics of a spring to the statistical mechanics of a high polymer will be of considerable utility in examining the mechanical response of a polymeric material. The main source of utility lies in the relative simplicity with which the analogous spring-mechanical problem may be treated.

II. VISCOSITY IN DILUTE SOLUTIONS

The behavior of *dilute* polymer solutions has received much attention because of the strong relationship of solution viscosity to polymer structure and configuration. In the discussions to follow it will be seen that polymer solutions are usually treated as two-phase systems consisting of mechanical elements (the polymer molecules) embedded in a continuum (the solvent). The solvent is imagined to exert forces upon the polymer in much the same way that a fluid exerts forces upon a small suspended particle. This model, although simple-minded and useful, is fundamentally improper. The interaction of solvent molecules with monomer elements cannot really be considered the same as the interaction of a continuous fluid with a solid since solvent molecules are not really microscopic in comparison with the monomer elements. Despite this fundamental objection the "two-phase" approach is well established and, indeed, has led to some remarkably successful and useful theories.

A. Viscosity of a Suspension of Spheres

It might be well to point out that even the behavior of suspended macroscopic solids in simple newtonian fluids is imperfectly understood. The classical problem in this area was formulated and solved by Einstein in 1905 (5). In that analysis he considered the perturbation to a simple shear flow due to the presence of a single solid sphere of neutral density. Because the flow was perturbed an excessive dissipation of energy occurred. This was viewed as equivalent to an increase in the viscosity of the suspension. By assuming that in a *dilute* suspension of spheres this single sphere calculation would still be valid for each individual sphere, Einstein arrived at an expression for the viscosity of a suspension relative to the viscosity of the suspending fluid. His result was

$$\eta/\eta_s = 1 + 2.5\phi \qquad (4.16)$$

where ϕ is the volume fraction of spheres.

Equation 4.16 has been verified for dilute suspensions, and it is generally recommended for ϕ up to 1% solids. For larger ϕ, interaction among particles becomes important, and the viscosity increases more rapidly than linearly with ϕ. The reader interested in the behavior of suspensions might read the review articles of Sadron (6) and of Frisch and Simha (7), which treat the behavior of concentrated suspensions and the behavior of

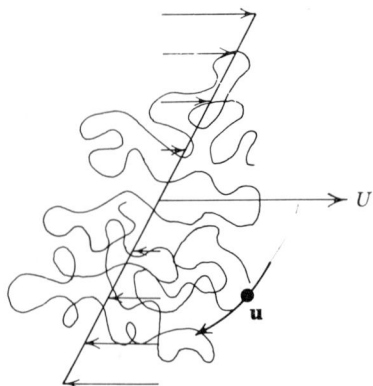

Fig. 4.4. Rotation of a polymer chain about its center of mass, due to the torque exerted by a shear field.

non-spherical particles. A recent critical analysis and correlation of experimental data on the viscosity of suspensions of spheres is given by Thomas (8).

While the behavior of suspensions is of considerable interest, its discussion here is of value only to the extent that it is pertinent to the calculation of the viscosity of a polymer solution. It will be seen that one can use the Einstein analysis for dilute polymer solutions, but the results leave much to be desired in terms of agreement with experiment. On the other hand, the general procedure of calculating the dissipation of energy due to the presence of a perturbing phase (the polymer molecules) is used in all the considerations of dilute solution behavior to be discussed here.

B. The Free Draining Polymer

One of the earliest quantitative approaches to the problem of predicting the viscosity of dilute polymer solutions is that of Debye (9). He considered an isolated polymer molecule in a simple shear field, as in Fig. 4.4. The velocity field may be taken as $\mathbf{U} = (\dot{\gamma}x_2, 0, 0)$, with the origin of the coordinate system fixed at the center of gravity of the molecule.* Debye made the following simplifying assumptions:

1. Individual monomer units do not interact with each other, nor do they distort the flow field. Such a molecule is called a "free draining coil."
2. Monomer inertia is ignored.

* Otherwise **U** may be considered to be the velocity of the fluid relative to the velocity of the center of gravity of the molecule.

3. If a monomer unit moves relative to the solvent, a frictional retardation is felt which is proportional to the relative velocity.

At equilibrium the center of gravity of the molecule will simply translate with the fluid velocity at that point. But the flow at points removed from the center of gravity will exert a moment on the molecule which will cause it to rotate about its center of gravity. Debye showed that the angular velocity ω would be exactly half the shear rate: $\omega = \frac{1}{2}\dot{\gamma}$.

A monomer group with coordinates (x_1, x_2) relative to the center of gravity will have a velocity, due to rotation, given by

$$\mathbf{u} = (\omega x_2, -\omega x_1, 0) \tag{4.17}$$

The relative velocity between this monomer and the solvent is

$$\mathbf{u}_r = \mathbf{U} - \mathbf{u} = \tfrac{1}{2}\dot{\gamma}(x_2, -x_1, 0) \tag{4.18}$$

If ζ is taken as the "monomer friction coefficient," the retarding force experienced by each monomer will be

$$\mathbf{F} = \zeta(\mathbf{U} - \mathbf{u}) \tag{4.19}$$

The work performed in overcoming this friction gives rise to an energy dissipation *per monomer* ϵ, where

$$\epsilon = \mathbf{F}\cdot(\mathbf{U} - \mathbf{u}) = \tfrac{1}{4}\zeta\dot{\gamma}^2(x_1^2 + x_2^2) \tag{4.20}$$

For the freely drained coil the total dissipation ε *per molecule* is simply

$$\varepsilon = \tfrac{1}{4}\zeta\dot{\gamma}^2 \sum_{i=1}^{N} (x_1^2 + x_2^2)_i \tag{4.21}$$

where the summation is extended over all monomer units of the molecule. Finally, if there are ν identical polymer molecules per unit volume, the volumetric dissipation rate is

$$\Phi = \tfrac{1}{4}\nu\zeta\dot{\gamma}^2 \sum_{i=1}^{N} (x_1^2 + x_2^2)_i \tag{4.22}$$

It is convenient at this point to introduce some macroscopic measure of molecular size. This may be done by defining an effective radius R_e as

$$R_e^2 = \frac{5}{2N} \sum_{i=1}^{N} (x_1^2 + x_2^2)_i \tag{4.23}$$

By analogy to the expression for dissipation in a homogeneous viscous fluid in simple shear, the viscosity *increase* due to the presence of polymer molecules may be calculated as $\eta_p = \Phi/\dot{\gamma}^2$, and the viscosity of the solution may be taken as the sum of the solvent viscosity η_s and the "polymer viscosity increment" η_p. In general, dilute solution viscosity is presented in terms of the "intrinsic viscosity," $[\eta]$, defined as

$$[\eta] = \lim_{c \to 0} \frac{\eta - \eta_s}{c\eta_s} \qquad (4.24)$$

where c is the concentration of polymer in grams per milliliter of solvent.* If c is introduced in place of ν, which has the effect of introducing the monomer molecular weight m, the intrinsic viscosity may be written, for the Debye theory, as

$$[\eta] = N_0 \zeta R_e^2 / 10 \eta_s m \qquad (4.25)$$

where N_0 is Avogadro's number. For a long linear polymer R_e^2 may be shown proportional to M and the intrinsic viscosity may be rewritten simply as

$$[\eta] = K_m M \qquad (4.26)$$

where K_m is a constant peculiar to the monomer, the viscosity of the solvent, and the frictional coefficient ζ. K_m will depend upon the configuration of the polymer insofar as that will affect both ζ and R_e^2. Equation 4.26 is known as Staudinger's viscosity rule.

C. Hydrodynamic Shielding

Deviations from the Staudinger rule, which are common, are generally attributable to the fact that a polymer molecule is not "freely drained." Intrinsic viscosities are usually found to follow

$$[\eta] = K_m M^\alpha \qquad (4.27)$$

with α taking values between 1.0 and 0.5. The concept of the freely drained coil, while appealing in its simplicity, is not very realistic. More sophisticated treatments of intrinsic viscosity take a more detailed view of the flow perturbation introduced by the monomer units. In particular, the idea of a "shielding effect" is introduced, whereby peripheral monomer units are imagined to be able to shield interior monomer units from the external flow.

Debye and Bueche (10) and Kirkwood and Riseman (11) have carried out such analyses which are similar in philosophy but different in mathematical technique. Each presents a mathematical treatment of the perturbation to the flow field interior and exterior to a polymer molecule. From their results the intrinsic viscosity is calculated and shown to follow Eq. 4.27. The exponent α is a function of a shielding parameter σ given by

* Note that $[\eta]$ has units of reciprocal concentration and not viscosity units. It is common to express $[\eta]$ in deciliters/gram, corresponding to a concentration unit of grams/deciliter (g./dl.).

Debye and Bueche as

$$\sigma^2 = 3N\zeta/4\pi R_e \eta_s \tag{4.28}$$

Kirkwood and Riseman obtain nearly the same numerical results as Debye and Bueche, and have a parameter which is equivalent to σ above. The differences between the two theories seem to be within experimental errors involved in measuring parameters like R_e and one cannot recommend one theory over the other on experimental grounds.

The major difference between the two theories lies in the model of the internal structure of a polymer molecule. Kirkwood and Riseman consider the polymer to be a Gaussian chain with a non-uniform spatial distribution of monomer units, more densely packed at the center of the molecule than on its periphery. Debye and Bueche deliberately ignore this complication and instead consider the polymer to be a porous sphere with a uniform distribution of monomer units. Since the differences between the two theories are small it would seem that the behavior of the molecule is fairly insensitive to the difference in assumed spatial distribution of monomer units.

The Debye-Bueche result may be written as

$$[\eta] = (4\pi R_e^3 N_0/3M)\varphi(\sigma) \tag{4.29}$$

where $\varphi(\sigma)$ is the function which accounts for the degree to which the peripheral monomer units of the polymer "sphere" are successful in shielding the inner monomer units from the external flow. The limiting behavior of φ may be shown to be

$$\varphi(0) = \sigma^2/10 \tag{4.30}$$

and

$$\varphi(\infty) = \tfrac{5}{2} \tag{4.31}$$

If Eq. 4.30 is substituted into Eq. 4.29, the result is equivalent to Eq. 4.25. Thus, when σ is small, the freely drained limit is approached. On the other hand, if Eq. 4.31 is substituted into Eq. 4.29, it may be shown that the result is identical with what one would obtain if Einstein's result (Eq. 4.16) were rewritten in terms of intrinsic viscosity. Thus, for large values of σ, the porous structure of the "sphere" essentially traps solvent within it and behaves like a rigid impenetrable sphere.

From the results presented by Debye and Bueche it is possible to prepare a plot of α and φ as functions of σ. From such a plot, as in Fig. 4.5, it is possible to calculate molecular dimensions of polymers in solution. Consider, for example, the data of Goldberg, Hohenstein, and Mark, quoted by Hermans (12). They found, for a solution of polystyrene in toluene, with polystyrenes having molecular weights in the range 2×10^5 to 18×10^5, that $K_m = 0.037$ and $\alpha = 0.62$. If one uses an average M of

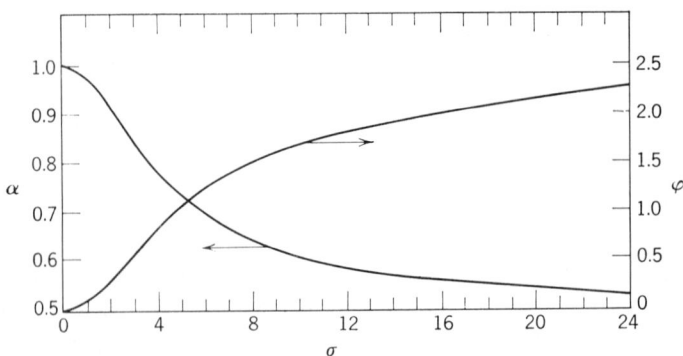

Fig. 4.5. Functions in the Debye-Bueche theory of hydrodynamic shielding.

10^6 in Eq. 4.27, a value of $[\eta]$ of 196 cm.^{3}/g. is found. From Fig. 4.5, for $\alpha = 0.62$, it is found that $\sigma = 9.2$ and $\varphi = 1.63$. From Eq. 4.29 R_e is found to be 500 Å. It is possible to show that $2R_e$ is nearly equal to the average end-to-end distance of a polymer chain, $\langle r^2 \rangle^{1/2}$ (10). It is possible to measure $\langle r^2 \rangle^{1/2}$ by a light scattering experiment (2, p. 115), and so one may estimate R_e. A number of such results are presented by Debye and Bueche (10), all of which indicate that R_e is in the range of 500 to 550 Å units, in agreement with this calculation.

While this is an encouraging result, it is by no means a strong confirmation of the theory. In the first place, one can find data which do not agree so well with the theory, and indeed Hermans quotes results with values of α greater than unity. Furthermore, this theory assumes that the polymer molecules are all of a single molecular weight; in practice this is not usually the case, and any confirmation of the theory is subject to assumptions concerning the proper way to account for a spectrum of molecular weights. It should also be apparent that, for a given set of data of $[\eta]$ vs. M, different values of α may be obtained by suitably adjusting the values of K_m. Hence, depending on the techniques used in curve-fitting the data, one would obtain different estimates of σ, and so of R_e. Finally, it is known that different results are obtained depending on the solvent used, although the theory does not account for differences in solvent properties. Solvent effects usually take the form of excluded volume effects which would have to be accounted for by modifying the coefficient in Eq. 4.3.

It seems fair to conclude that the Debye-Bueche theory has met with some confirmation, particularly in its ability to account for the variation in α observed in most polymer systems. This would tend to confirm the basic physical ideas involved in accounting for monomer interaction and flow perturbation within the polymer molecule. The failures of the theory must be ascribed, as usual, to the introduction of simplifications which cannot be supported, but without which no quantitative predictions would result.

D. The Theta-Solvent

In the limit of large σ Eq. 4.29 reduces to

$$[\eta] = K_\Theta M^{1/2} \tag{4.32}$$

For most linear polymers it is possible to find a solvent such that Eq. 4.32 is obeyed at some particular temperature. Such a solvent is known as a Θ (theta)-solvent, and the temperature at which the $\log [\eta] - \log M$ curve has a slope of one-half is known as the Θ-temperature. This behavior is associated with the action of a poor solvent, in which the polymer tends to reduce its contact with the solvent, and so adopt a tightly coiled configuration free of excluded volume effects. Values of the Θ-temperature and K_Θ are tabulated for a variety of polymer-solvent systems (13,14).

Figure 4.6 shows intrinsic viscosity behavior of polystyrene in media of varying solvent power (15). Cyclohexanone is a Θ-solvent at 34.5°C. Butanone and toluene are increasingly better solvents for polystyrene and the $[\eta] = M$ curve reflects this change. At low molecular weights the solvent effect seems to vanish. This is interpreted as a decrease in excluded

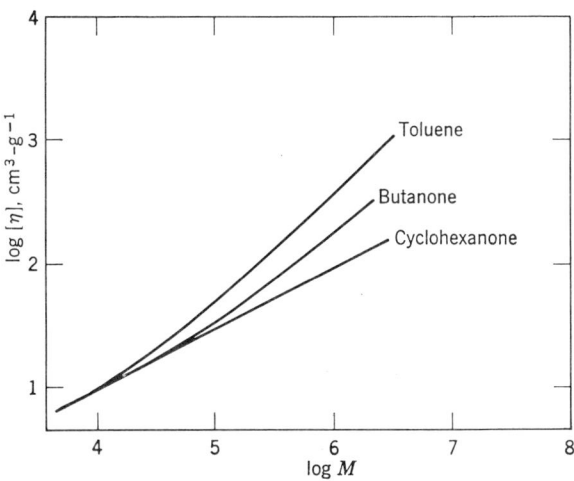

Fig. 4.6. Intrinsic viscosity of polystyrene in solvents of varying power.

volume effect as molecular weight decreases, as long as the molecular weight does not become so small that the polymer can no longer be considered a Gaussian chain (15).

E. Effect of Polydispersity

Equation 4.27 may be used for the determination of molecular weight through the measurement of intrinsic viscosity. The coefficients K_m and α must first be established for a series of known molecular weight samples in a particular solvent, at some fixed temperature. In view of the temperature-solvent-molecular weight dependence of α, it is advisable to measure the intrinsic viscosity of the unknown molecular weight sample under conditions closely approximating those for which K_m and α were determined.

If the polymer is polydisperse it is still possible to determine its molecular weight, although one now obtains some molecular weight average. At low concentrations one would expect a good approximation to $[\eta]$ to be

$$[\eta] = \lim_{c \to 0} \frac{\eta - \eta_s}{c\eta_s} \approx \frac{\eta - \eta_s}{c\eta_s} \tag{4.33}$$

or

$$\eta - \eta_s \approx c\eta_s[\eta]$$

For a dilute polydisperse sample one would expect the viscosity contribution from each molecular weight fraction to be additive, so that

$$\eta - \eta_s = \eta_s \sum_i c_i[\eta]_i \tag{4.34}$$

If each fraction follows $[\eta]_i = K_m M_i^\alpha$ then

$$\eta - \eta_s = \eta_s \sum_i c_i K_m M_i^\alpha \tag{4.35}$$

If division by $c\eta_s$ is performed and the limit examined, the result is

$$\lim_{c \to 0} \frac{\eta - \eta_s}{c\eta_s} = [\eta] = K_m \sum_i \left(\frac{c_i}{c}\right) M_i^\alpha = K_m M_v^\alpha \tag{4.36}$$

where a "viscosity average" molecular weight has now been defined by $[\eta] = K_m M_v^\alpha$. By inspection, then, it is seen that

$$M_v = \left[\sum_i w_i M_i^\alpha\right]^{1/\alpha} = ([\eta]/K_m)^{1/\alpha} \tag{4.37}$$

where w_i is the weight fraction of each molecular species.

In view of previous remarks it should be clear that Eq. 4.37 is on somewhat shaky grounds since K_m and α may depend on molecular weight. This is a particular problem if the polymer examined has a very broad molecular weight distribution. Meaningful results may be obtained, however, if K_m and α are established for polymers of known molecular weight distribution, and if the sample has a similar distribution. It is known that M_v lies between M_n and M_w and is always closer to M_w (1, p. 313).

III. NON-NEWTONIAN VISCOSITY

A. The Bueche Theory of Viscosity (16,17)

All the analyses of dilute solution polymer viscosity described above omit consideration of the effect of shear rate on the response of the molecule. For this reason it should be understood that the ensuing viscosity results are valid only at low shear rates. Since it is not possible to estimate *a priori* with complete assurance just what the shear rate restrictions might be for a given polymer solution, it is advisable to perform viscosity measurements at more than one shear rate. If no discernible effect of shear rate is found, one may conclude that the results represent "zero shear" values.

An analysis of the dynamic behavior of a polymer molecule in a shear field was carried out with some success by Bueche (16,17). His starting point was the Debye picture of the free draining coil rotating in a simple shear field. As described earlier (p. 141), this molecule is supposed to rotate with a frequency of one-half the shear rate. With reference to Fig. 4.7, it can be seen that the monomer units along the line AA' experience a

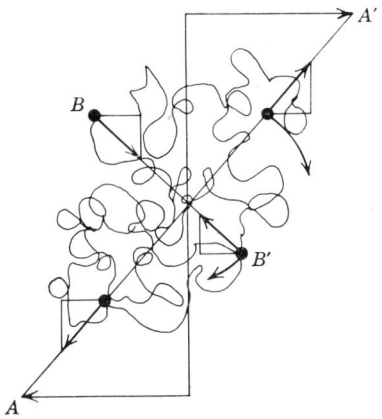

Fig. 4.7. Forces exerted on a monomer as it rotates about the center of mass of the polymer. The monomer experiences periodic tension and compression as it passes across the axes AA' and BB'.

drag force which tends to extend the chain. Along the line BB', however, the drag force tends to compress the chain. Hence, as a monomer unit rotates about the center of gravity of the molecule, it experiences a sinusoidally alternating tension and compression.

The formal correspondence of the effect of a force acting on a Gaussian chain polymer molecule to the force acting on a simple spring has already been established (Eq. 4.15). Bueche imagined the polymer molecule to be subdivided into a large number of "sub-molecules," each of which behaves mathematically like a small mass attached to a linear spring. The dynamics of the polymer molecule is thereby replaced by the dynamics of a set of masses connected in series by elastic springs. This latter problem, although complex, allows the development of a formal solution for the displacement of each sub-molecule relative to its equilibrium position. In the case of a periodic forcing function, arising from the rotation of the molecule in the shear field, the displacements of the monomer units are also periodic. Once the displacements are known for each sub-molecule the dissipation may be calculated, as in Debye's analysis, as the product of force and velocity (time rate of change of displacement) relative to the solvent. When this quantity is added over the entire molecule, and further averaged over a period of rotation, the viscosity follows directly. Bueche's result takes the form

$$\frac{\eta - \eta_s}{\eta_0 - \eta_s} = 1 - \frac{6}{\pi^2} \sum_{n=1}^{N} \frac{(\dot{\gamma}\lambda_1)^2}{n^2(n^4 + \dot{\gamma}^2\lambda_1^2)} \left(2 - \frac{(\dot{\gamma}\lambda_1)^2}{n^4 + \dot{\gamma}^2\lambda_1^2}\right) \quad (4.38)$$

where η_0 is the viscosity in the limit of zero shear rate.

The parameter λ_1 is a relaxation time, given by Bueche's analysis as

$$\lambda_1 = 12(\eta_0 - \eta_s)M/\pi^2 cRT \quad (4.39)$$

The analysis can be extended to an undiluted polymer, and the results are identical to the above, with the solvent viscosity η_s set equal to zero, and with c replaced by the polymer density ρ. If η_0 is measured, and concentration c, molecular weight M, and absolute temperature T are known, one may predict the $\eta(\dot{\gamma})$ curve without the introduction of any adjustable parameters. An example of the degree to which Bueche's theory corresponds with observed behavior is given in Fig. 4.8. Considering the lack of adjustable constants, one must agree that the theory predicts the general features of non-newtonian behavior with fair accuracy. More detailed comparisons will be given in Chapter 5.

According to Eq. 4.38 the "reduced viscosity," $(\eta - \eta_s)/(\eta_0 - \eta_s)$, is a function only of $\dot{\gamma}\lambda_1$. For a given material, that is, at fixed values of c (or ρ) and M, λ_1 is a function only of temperature. Hence the reduced

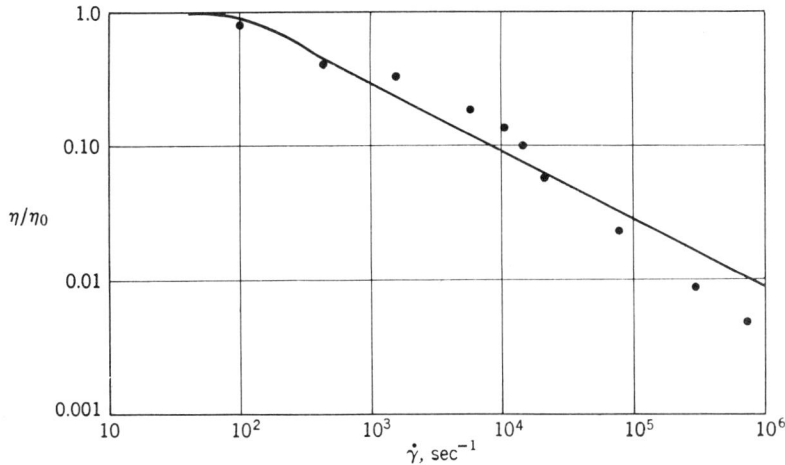

Fig. 4.8. A test of Bueche's theory, for molten polydimethylsiloxane (Union Carbide L-45 silicone oil, at 25°C.). Data taken from G. C. Johnson, *J. Chem. Eng. Data*, **6**, 275 (1961). Viscosity average molecular weight given as $M_v = 9.5 \times 10^4$ by E. A. Collins and W. H. Bauer, *Trans. Soc. Rheol.*, **9:2**, 1 (1965). The data, obtained using short capillaries, were corrected (by Johnson) for entrance losses, but not for non-newtonian effects. If the Weissenberg-Rabinowitsch-Mooney equation (Eq. 2.12) were used, the data points at high $\dot{\gamma}$ would shift to the right by a factor of approximately 1.4.

viscosity is affected by shear rate in essentially the same way as it is affected by a shift in temperature. One could define a "shift factor" a_T as

$$a_T = (\eta_0 - \eta_s)_T T_0 / (\eta_0 - \eta_s)_{T_0} T \tag{4.40}$$

where T_0 would be some arbitrary reference temperature. Data taken at different temperatures, when plotted as reduced viscosity against $a_T \dot{\gamma}$, would be identical with a plot of reduced viscosity against $\dot{\gamma}$ alone, at the reference temperature T_0. If this is so, one could use data obtained at different temperatures to predict viscosity at shear rates which might not be accessible on a particular viscometric system. Figure 4.9 gives an example of data obtained at various temperatures and Fig. 4.10 shows the superposition of these data to a single curve, as predicted by Eqs. 4.38 and 4.39.

In Figures 4.11 and 4.12 concentration superposition is tested, while Figs. 4.13 and 4.14 give an example of molecular weight superposition. In all these examples it may be seen that the Bueche theory is not quantitatively exact. The more general prediction that the reduced viscosity is a function of $\dot{\gamma}\lambda_1$ seems to be fairly well borne out, however.

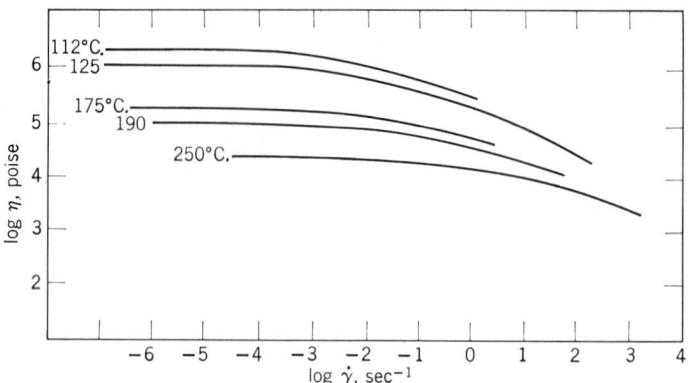

Fig. 4.9. Molten polyethylene data at various temperatures. (W. Philippoff and F. Gaskins, *J. Polymer Sci.*, **21**, 205 (1956)). Data were obtained in a coaxial cylinder instrument and were corrected for non-newtonian flow using Eq. 2.49.

B. Extension to Polydisperse Polymers (18)

Bueche's analysis assumes that the polymer is monodisperse. The question naturally arises as to whether the theory is applicable to a polydisperse material, and, if it is, to what molecular weight average should the relaxation time λ_1 be referred.

Experimental evidence indicates that the *shape* of the η vs. $\dot{\gamma}$ curve is altered by polydispersity. Figure 4.15 shows data for two polyisobutylenes of different molecular weight distributions. The polydisperse material deviates from newtonian behavior at a lower shear rate than the more

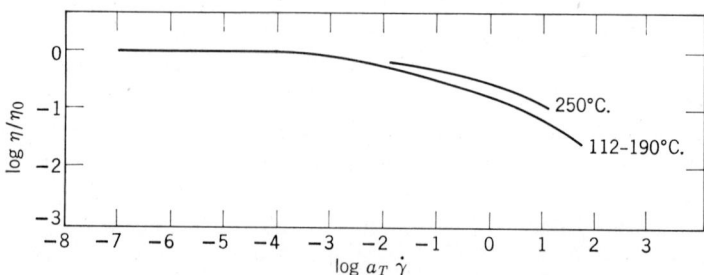

Fig. 4.10. Data of Fig. 4.9 "shifted" to an arbitrary reference temperature of 112°C. Vertical shift for each set of data accomplished using η_0 at the temperature corresponding to that set of data. Horizontal shift uses Eq. 4.40 for a_T, T_0 being taken as 112°C. (385°K.). Only the data farthest removed from the reference temperature fail to shift into acceptable coincidence.

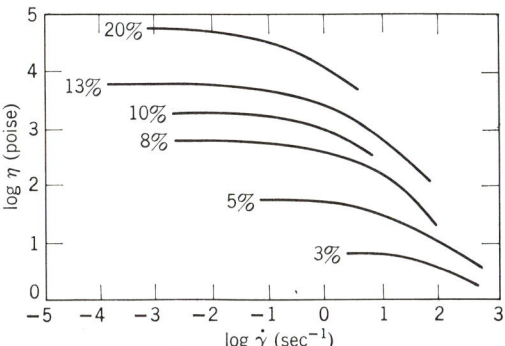

Fig. 4.11. Viscosity data for solutions of a high molecular weight polyisobutylene (PIB) in decalin at 25°C. obtained with a cone and plate viscometer (T. W. DeWitt et al., *J. Colloid Sci.*, **10**, 174 (1955). The molecular weight averages of the polymer are given as $M_n = 3.7 \times 10^5$ and $M_w = 1.06 \times 10^6$.

narrow distribution polymer, and shows a more gradual rate of change of viscosity with shear rate. Such behavior is typical and has been observed in other polymers (19, 20). At best, then, one might use the Bueche theory by inserting in λ_1 *some* average molecular weight which would minimize the deviation between theory and experiment. Experimental evidence indicates some success if the viscosity average or weight average molecular weight is used (21). One can also find recommendations for the

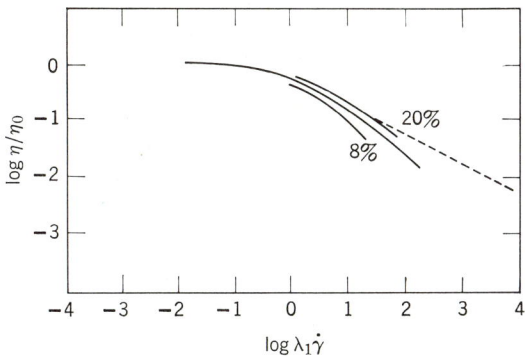

Fig. 4.12. Data of Fig. 4.11 replotted as a test of the Bueche theory, with λ_1 calculated from Eq. 4.39. The dashed line is the Bueche theory (Eq. 4.38). Concentration superposition is good except for the 8 and 20% data, which fall on either side of the rest of the data of Fig. 4.11.

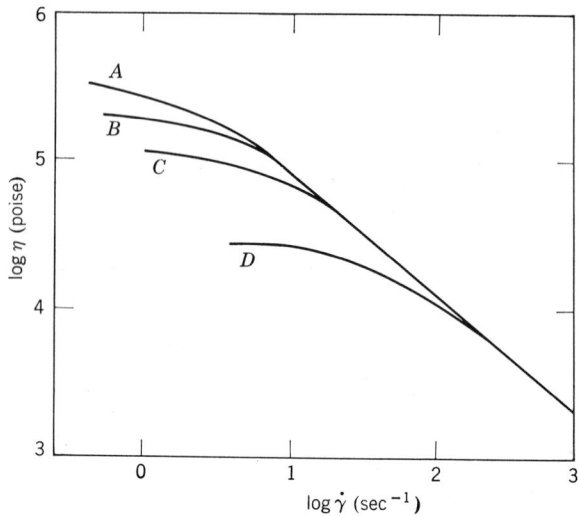

Fig. 4.13. Viscosity data obtained at 183°C. with four monodisperse polystyrenes (26). Non-newtonian capillary flow correction factor (Eq. 2.12) was as large as 2.2. No correction was made for entrance effects, which were alleged to be negligible, although the largest L/D was 20. The ratio M_w/M_n is less than 1.1.

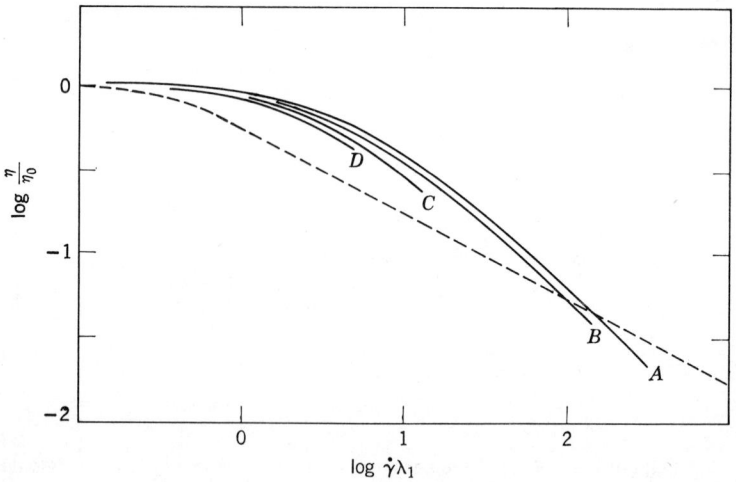

Fig. 4.14. Data of Fig. 4.13 replotted as a test of the Bueche theory, shown as the dashed line. Note the failure of the data to superimpose.

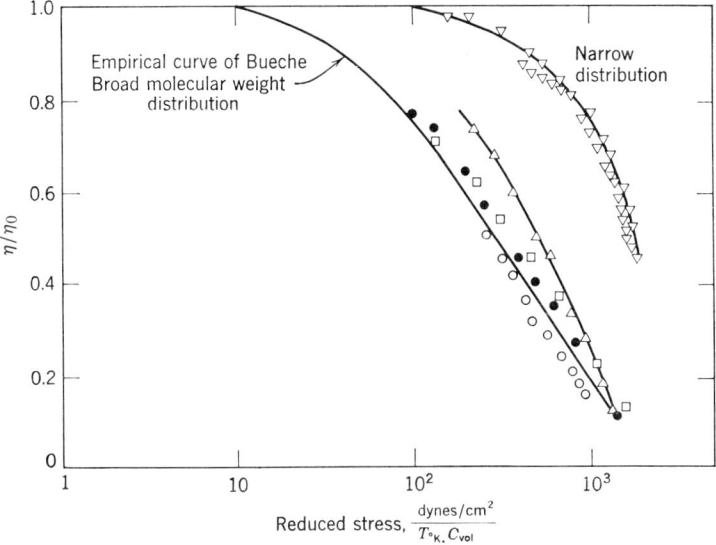

Fig. 4.15. Viscosity of polyisobutylenes of different molecular weight distributions (19).

"z-average" (22)

$$M_z = \int_0^\infty M^3 f(M)\, dM \Big/ \int_0^\infty M^2 f(M)\, dM \qquad (4.41)$$

and evidence that different averages are appropriate to the low and high shear rate regions (23).

This seems an unsatisfactory situation, and so an attempt was made to modify Bueche's theory to be applicable to a polydisperse material (18). The main idea of the modification is that Bueche's theory is assumed applicable to each fraction of monodisperse polymer. The contribution to the viscosity from each fraction is then summed according to the distribution of each fraction.

If the right-hand side of Eq. 4.38 is considered a function of molecular weight, say $F(M)$, then Bueche's theory becomes (18)

$$(\eta - \eta_s)/(\eta_0 - \eta_s) = \int_0^\infty M^2 F(M) f(M)\, dM / M_w M_n \qquad (4.42)$$

and the resultant curve depends upon the distribution function $f(M)$. It is then possible to plot the reduced viscosity against $\dot{\gamma}\langle\lambda_1\rangle$, where $\langle\lambda_1\rangle$ is a relaxation time based upon some particular average molecular weight. Figure 4.16 shows such a curve based upon the number average M_n.

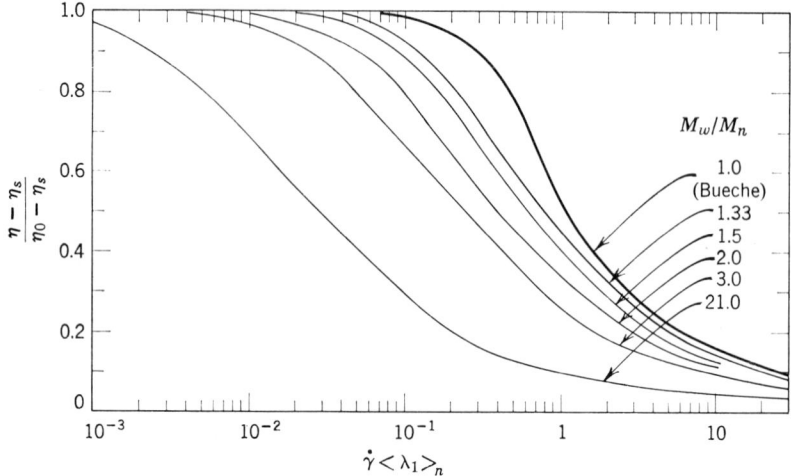

Fig. 4.16. Theoretical viscosity curves for polydisperse materials (18). Relaxation time based on the number average molecular weight.

The figure shows qualitative agreement with the observations concerning the early departure from newtonian flow and the reduced rate of viscosity reduction for the broad distribution polymers.

In Fig. 4.16 $f(M)$ was taken as the "most probable distribution," defined by

$$f(M) = B^A M^{A-1} e^{-BM}/\Gamma(A) \qquad (4.43)$$

where A and B are adjustable constants and $\Gamma(A)$ is the "gamma function." It may be shown that

$$M_n = A/B \qquad (4.44)$$

and

$$M_w/M_n = (A - 1)/A \qquad (4.45)$$

Hence A and B may be established from two molecular weight determinations, M_n and M_w, assuming the distribution is as given in Eq. 4.43. If the molecular weight distribution function is known, but is not expressible in some simple analytical form, it is still possible to calculate the reduced viscosity by numerical integration.

Unfortunately there are very few data in the literature which offer a complete $\eta(\dot\gamma)$ curve along with a complete characterization of the molecular weight distribution. One such study was carried out by Porter, Cantow, and Johnson (24) for solutions of polyisobutylene. Their results have been replotted in Fig. 4.17. It can be seen that the modification of the Bueche theory offers considerable improvement over the unmodified

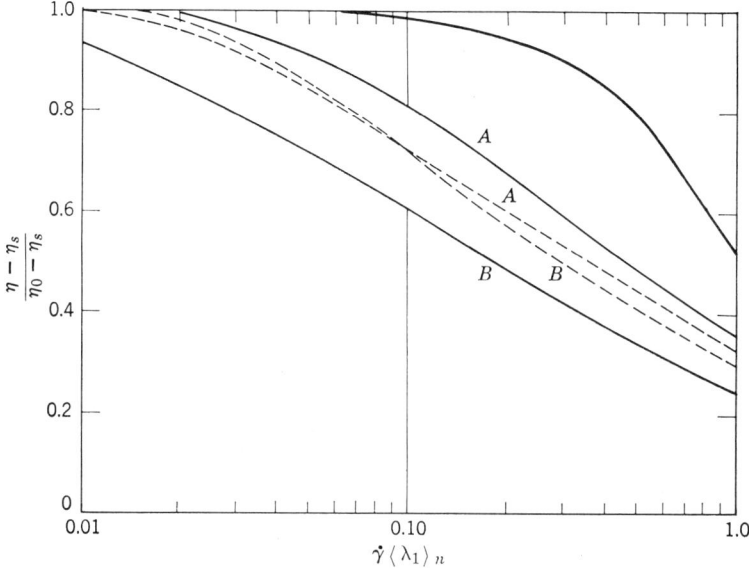

Fig. 4.17. Test of the polydisperse theory with data for polyisobutylene solutions. Bueche's theory (shown as the upper line) used as a basis for the polydisperse theory (shown as the lower dashed lines). The lower full lines are data from Ref. 24.

monodisperse theory, but there still is not strict quantitative agreement with experiment. In view of the simplicity of Bueche's model, however, and its lack of adjustable parameters, it must be viewed as a very useful theory if one has no other information available.

The most serious drawback of the polydisperse theory, as formulated above, is that it is based on Bueche's theory. In the absence of monodisperse data, however, one has no choice but to make some assumption for the dependence of viscosity on molecular weight, and on the other parameters that appear in a group such as $\dot{\gamma}\lambda_1$. Robinson (25) has suggested an approach which is quite successful, although it requires extensive data on monodisperse materials.

Robinson uses Eq. 4.42, but obtains $F(M)$ not from Eq. 4.38, but by curve-fitting *data* for monodisperse samples of the polymer of interest. He tested this approach by examining the viscosity–shear rate behavior of a series of blends of two commercial polystyrenes. Molecular weight distribution curves were determined for the two polystyrenes, and so could be calculated for the blends of the two materials. The viscosity data of Stratton (26) and of Ballman and Simon (27) for monodisperse polystyrenes were replotted and curve-fit by a least squares computer routine.

In place of Eq. 4.38, $F(M)$ was found to be

$$F(M) = \eta/\eta_0 = [1 + 2.68(\dot{\gamma}\lambda_1)^{0.89}]^{-1} \qquad (4.38a)$$

The relaxation time λ_1 was given, not by Eq. 4.39, but by

$$\lambda_1 = 12\eta_0 M^{0.75}/\pi^2 \rho RT \qquad (4.39a)$$

Stratton's observation, with which Robinson concurs, is that λ_1 depends on a power of M lower than first power.

Following the details of the polydisperse theory (18), then, λ_1 in Eq. 4.38a is replaced by

$$\lambda_1 = M^2 \langle \lambda_1 \rangle_w / M_w^2 = (M^2/M_w^2)(12\eta_0 M_w^{0.75}/\pi^2 \rho RT)$$

Then $F(M)$ and $f(M)$ are introduced into Eq. 4.42, and one may prepare curves of η/η_0 as a function of $\dot{\gamma}$, or $\dot{\gamma}\langle\lambda_1\rangle_w$. Robinson presents such curves for polydisperse samples of molten polystyrene. Experimental data for these same materials were then obtained with an Instron capillary rheometer. The agreement between theory and experiment is quite good, as can be seen in Fig. 4.18.

C. Molecular Entanglement Theory

The idea of polymer molecules, or segments of molecules, undergoing unhindered steady rotation in concentrated solution, or in the molten

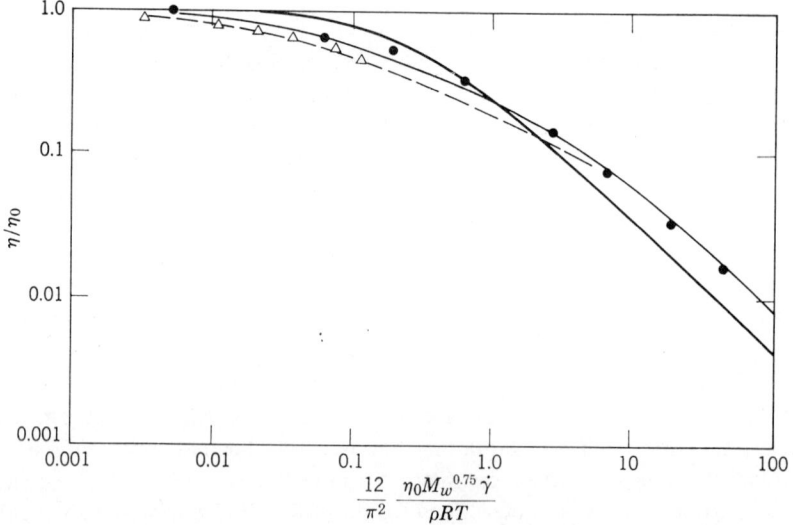

Fig. 4.18. Test of polydisperse theory with data for molten polystyrene. Monodisperse data (represented by heavy curve) used as a basis for the theory. Polydisperse theory and data ($\bullet$), $M_w/M_n = 2.0$. Polydisperse theory and data ($\triangle$), $M_w/M_n = 4.7$.

state, is very difficult to accept. Graessley has offered an alternative picture of polymer behavior which leads to a non-newtonian viscosity without requiring rotation (29). He envisions events between molecular segments during which interaction, which he considers to be entanglement, occurs leading to increased dissipation of energy.

The entanglement process is a complicated kinetic phenomenon. Two molecules may be thought of as approaching one another in a shear field. When they are sufficiently close, entanglements begin to occur at a finite rate. As the molecules pass, disentanglement occurs. An entanglement density for the bulk material may be defined to characterize the number of entanglements which exist at any instant, averaged over the material.

The kinetics of such a process, of course, is unknown. Graessley adopted a very simple mechanistic picture. For an entanglement to exist, two molecules must first be within a certain distance of each other, say within a sphere of radius R. Second, the molecules must remain within this sphere for a finite time λ, or else no entanglement occurs. Figure 4.19 pictures this process. Now, the greater the shear rate, the more rapidly two molecules move relative to one another. Hence, the entanglement density is reduced by high shear rate, since fewer molecules will remain in the "entanglement sphere" for a sufficiently long time at high shear rate.

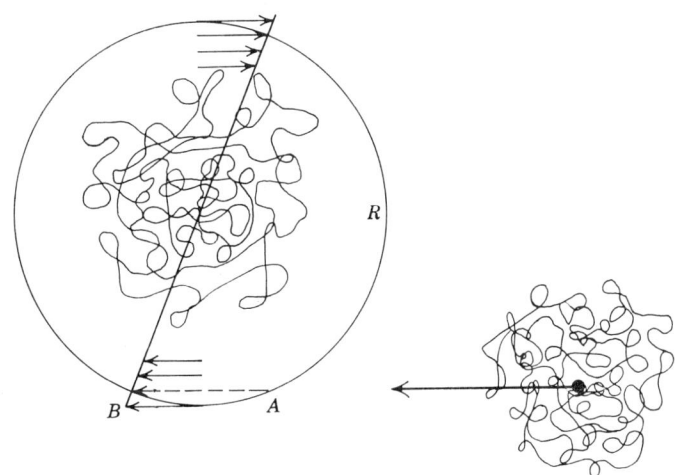

Fig. 4.19. Definition sketch for Graessley's entanglement theory. The polymers move relative to one another because of a velocity gradient. If the center of mass of one molecule moves through the sphere centered about the second molecule, entanglement may occur. The duration of the entanglement depends upon the time required to move along the path AB, and is reduced as the velocity gradient increases.

When the details of this dynamic picture are worked out, the viscosity may be calculated, and the result is $\eta/\eta_0 = F(\lambda\dot\gamma)$ just as in the Bueche theory. The zero shear viscosity contains certain unknown parameters of the model, such as the frictional coefficient associated with entanglement, the equilibrium entanglement density at zero shear rate, and the entanglement sphere radius. The time parameter λ is presumed to be proportional to Bueche's λ_1, with the constant of proportionality in the neighborhood of unity. The functional dependence of $F(\lambda\dot\gamma)$ is very much like that of the Bueche theory. For very large $\lambda\dot\gamma$ Graessley shows that $\eta/\eta_0 \sim (\lambda\dot\gamma)^{-3/4}$, while Bueche's theory gives $\eta/\eta_0 \sim (\lambda\dot\gamma)^{-1/2}$.

A calculation of the elastic energy associated with entanglements leads to an expression for the normal stress coefficient ψ_{12}, with the result $\psi_{12} = \psi_{12}^0 (\eta/\eta_0)^2$. While very few data exist for comparison (see Fig. 5.12), it is known that ψ_{12} falls off with $\dot\gamma$ more rapidly than does η, as predicted by the expression above. One concludes, from Graessley's analysis, that a consideration of the kinetics of molecular entanglement leads to a model which exhibits non-newtonian viscosity and shear-dependent normal stresses. Rotation of polymer molecules in a shear field is not prerequisite to such behavior, as implied in the Bueche theory.

In a subsequent experimental study of concentrated solutions of monodisperse polystyrene, Graessley and coworkers (28) find that viscosity data correlate in a manner generally consistent with the entanglement theory. The relaxation time, however, shows a molecular weight dependence which is somewhat weaker than first power. This is consistent with Stratton's observation in molten polymers.

D. The Molecular Theory of Williams

The molecular theories of Bueche and of Graessley both involve specific notions of a physical mechanism whereby viscous dissipation becomes shear rate dependent. Bueche considers individual molecules to be subject to rotation at a frequency proportional to the shear rate in the bulk material. The molecule is alternately stretched and compressed, and the dissipation of energy associated with this process is found to be dependent upon the frequency of this deformation and hence on the shear rate. In Graessley's development, molecular entanglement is taken to be the major source of frictional resistance to flow. A simple geometrical argument gives the density of entanglements as a decreasing function of the relative motion of molecules, and hence a function of shear rate.

A more general approach to a molecular theory is given by Williams (30), who makes no commitment to a specific mechanism of interaction between molecules, or between segments of individual molecules. Instead, it is assumed that an interaction does exist, and that the force of interaction can be expressed in terms of a *potential function* and a *segment distribution function*. The potential function is to be thought of as related to forces exerted on a given monomer segment by other monomer segments. Without specifying the physical nature of this force in any detail, it is assumed that the force depends upon some interaction coefficient and upon the spatial distribution of monomer segments surrounding that segment for which the force is being computed.* The segment distribution function describes this monomer segment spatial distribution.

At equilibrium, that is, at vanishing shear rates, the segment distribution function would be calculated by methods similar to those that lead, for example, to the chain-end distribution function discussed on p. 137. This would be a spherically symmetric, Gaussian, distribution function. Williams assumed that in a shear field a molecule would elongate in the direction of shear, and the corresponding segment distribution function would be skewed to a degree which depended upon the magnitude of the shear rate. An explicit, although empirical, form for the segment distribution function was taken, and in this way the shear rate was introduced explicitly into the model for interaction.

The result of Williams' analysis may be expressed in the familiar form $(\eta - \eta_s)/(\eta_0 - \eta_s) = F(\lambda \dot{\gamma})$. The zero shear viscosity is given explicitly in terms of parameters such as the interaction coefficient for segmental forces. The parameter λ is a scaling factor, with the dimensions of a time, which expresses the degree to which shear rate is effective in elongating a molecule, and thereby skewing the segment distribution function. The functional behavior of $F(\lambda \dot{\gamma})$ is qualitatively similar to that of other molecular theories. At high shear rates η/η_0 is predicted by Williams' theory to decrease like $1/\lambda \dot{\gamma}$, a much more rapid decrease than is predicted by other theories, or observed experimentally.

In a companion paper (32), Williams attempts a more detailed analysis of his model, and derives dependencies of η_0 and λ on measurable molecular and solution properties. He finds the zero shear viscosity to depend on concentration c and molecular weight M according to

$$\eta_0 \sim (cM^{0.625})^2 F^{1.5}(c)$$

where $F(c)$ is expected to be a weak function of c. The dependencies on c and M are considerably weaker than those usually observed in polymer solutions which exhibit significant non-newtonian effects, but the prediction that η_0 depends on the product $cM^{0.625}$ is in very good agreement with certain experimental observations (see Chapter 5).

* Although no direct *physical* interpretation is given to the origin of this force, its *mathematical* form is taken from an equilibrium theory of polymer solutions due to Fixman (31).

The time constant λ is found to be of the form $\lambda \sim (\eta_0 - \eta_s)/c^2 TF(c)$ in contrast to the Bueche theory, which gives $\lambda_1 \sim M(\eta_0 - \eta_s)/cT$. Attempts to correlate solution viscosity data over a wide range of concentrations lend some confirmation to $\lambda \sim 1/c^2$, rather than $\lambda_1 \sim 1/c$, but insufficient data exist to offer stronger support than this. The Bueche theory seems to be more realistic than the Williams theory with regard to the dependence of λ on M. The concentration and molecular weight superposition of η_0 and η/η_0 data is discussed more completely in Chapter 5.

A third study by Williams (33) leads to calculation of normal stress coefficients. ψ_{12}^0 is predicted to vanish in the limit of zero shear, clearly in contradiction of experimental results. ψ_{13}^0 is predicted to vary with $\lambda(\eta_0 - \eta_s)$, and is predicted to be negative (the stress difference corresponding to ψ_{13}^0 is thus a tension). If the λ dependence is introduced, this gives

$$\psi_{13}^0 \sim (\eta_0 - \eta_s)^2/c^2 TF(c)$$

Insufficient data are available to offer a proper test of this result.

As the shear rate increases, ψ_{12} and ψ_{13} decrease in magnitude like $1/(\lambda\dot{\gamma})^2$, ψ_{12} being slightly larger in absolute value. The difference ψ_{23}, which would vanish if the Weissenberg hypothesis were valid, is found to be positive and about 20% of the magnitude of ψ_{12}. Experiments indicate that ψ_{23} does become quite small with increasing shear rate (34).

One must conclude that Williams' analysis has much to recommend it. In particular it does not require a physical argument based on the notion of rotating molecules, as in the Bueche theory, or on the kinetics of entanglements, as in Graessley's work. Some of the detailed predictions are clearly at variance with experience, particularly the weak dependence of η_0 on molecular weight, and the too rapid fall-off of η/η_0 with $\dot{\gamma}$. These failures can in part be traced to simplifications in the details of relating parameters like the relaxation time λ to molecular structure, and to the empirical choice by which the segment distribution function is distorted by shear rate.

IV. DYNAMIC VISCOSITY

A. Zimm's Theory (35)

Zimm developed a theory of dynamic viscosity based upon Bueche's Gaussian sub-chain model of the polymer molecule. The molecule is imagined to be suspended in a viscous medium which offers hydrodynamic interaction calculated according to the method of Kirkwood and Riseman (11). In addition the molecule is considered to be subject to the effects of Brownian motion. A dynamic problem is then formulated for the case of an external flow which is oscillatory simple shear, so that $\dot{\gamma} = \text{Re}\{\gamma_0 e^{i\omega t}\}$.

The result for the dynamic viscosity may be written as

$$\eta^* = \eta_s + \frac{cRT}{M} \sum_{p=1}^{N} \frac{\lambda_1'}{p^2 + i\bar{\omega}\lambda_1'} \quad (4.46)$$

where λ_1' is called by Zimm the "terminal relaxation time," and is given as

$$\lambda_1' = \langle r^2 \rangle \zeta / 6kTq_1 \quad (4.47)$$

q_1 is the smallest of a set of eigenvalues obtained from an eigenvalue equation which cannot be solved exactly.* For this reason Zimm examined two limiting cases, corresponding to vanishing hydrodynamic interaction, (the freely drained coil) and dominant hydrodynamic interaction. The parameter which controls the choice of approximation is essentially the same as the parameter σ of the Debye-Bueche (or Kirkwood-Riseman) analysis (Eq. 4.28).

For vanishing hydrodynamic interaction Zimm finds

$$q_p = \pi^2 p^2 / N^2 \quad (4.48)$$

so that

$$\lambda_1' = \langle r^2 \rangle \zeta N^2 / 6\pi^2 kT \quad (4.49)$$

By assuming that Eq. 4.46 holds in the limit of vanishing concentration c and frequency $\bar{\omega}$, one may obtain an expression for the intrinsic viscosity at zero shear rate $[\eta]_0$ with the result

$$[\eta]_0 = \frac{RT\lambda_1'}{M\eta_s} \sum_{p=1}^{N} \frac{1}{p^2} \quad (4.50)$$

From this expression λ_1' may be written in terms of macroscopically measurable quantities to give

$$\lambda_1' = 6M\eta_s[\eta_0]/\pi^2 RT \quad (4.51)\dagger$$

If Eq. 4.46 is assumed to hold at any concentration then it follows that

$$\lambda_1' = 6M(\eta_0 - \eta_s)/\pi^2 cRT \quad (4.52)$$

and λ_1' is seen to be identical with the relaxation time calculated by Bueche (Eq. 4.39), except for a factor of 2.

This free draining limit is identical to a result developed by Rouse (36)

---

* The eigenvalue equation is essentially the dynamic equation for the model of the polymer chain set up by Zimm. The eigenvalues are just a particular set of constants which allow the solution to satisfy the boundary conditions on the eigenvalue equation.

† It may be shown that, for sufficiently large N, $\sum_{p=1}^{N} 1/p^2 = \pi^2/6$.

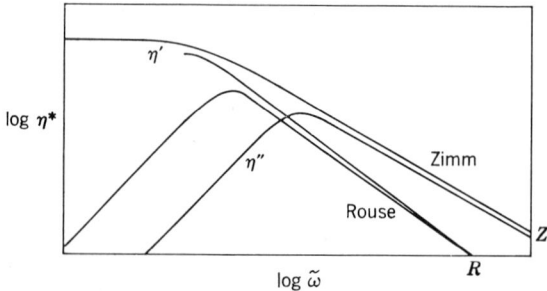

Fig. 4.20. Qualitative comparison of the components of η^* according to the Rouse or Zimm theories.

and will be referred to as the Rouse theory. Zimm's analysis was novel in presenting a theory in the limit of *dominant* hydrodynamic interaction. For that case Eq. 4.46 still holds, but λ'_1 is now given by

$$\lambda'_1 = M(\eta_0 - \eta_s)/2.369\, cRT \qquad (4.53)$$

A comparison of the Rouse and Zimm theories is shown in Fig. 4.20. In Fig. 4.21 the same curves are replotted in terms of the complex modulus $G^* = \eta^*/i\tilde{\omega}$, since this quantity, rather than η^*, is often calculated directly from experimental data.

B. Tschoegl's Extension

A comparison of experimental data with the theories of Rouse and Zimm indicates that data may be found in support of either theory (37), although most of the experimental results tend to support the Zimm theory more

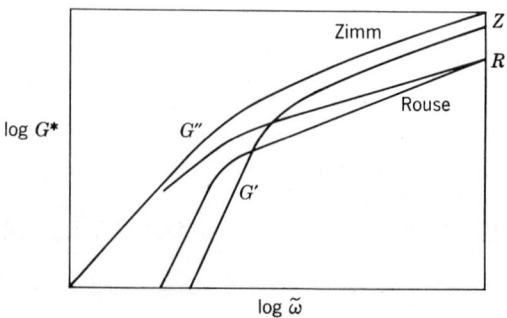

Fig. 4.21. Qualitative comparison of the components of G^* according to the Rouse or Zimm theories.

strongly than the Rouse theory. In an effort to bridge the gap between these two theories, Tschoegl has extended Zimm's treatment to account for intermediate values of hydrodynamic interaction (38). As expected, the resultant theory predicts curves for the components of η^* (or G^*) intermediate between the Rouse and Zimm curves. A shielding parameter h (related to the Debye-Bueche shielding parameter σ of Eq. 4.28 by $h = 4\sqrt{3}\,\sigma^2/9\sqrt{\pi}$) appears implicitly in Tschoegl's solution. η^* is still given by Eq. 4.46, but λ_1' is now calculated from

$$\lambda_1' = (\eta_0 - \eta_s)M/t(h)cRT \tag{4.54}$$

where $t(h)$ is a function of h which varies from $\pi^2/6$ to 2.369 as h varies from zero (free draining coil–Rouse) to infinity (dominant interaction–Zimm).

In a companion paper Tschoegl (39) extended a treatment by Ptitsyn and Eisner (40) which had accounted for the effect of excluded volume on the hydrodynamic interaction in the Zimm theory. As a result Tschoegl was able to modify the Zimm theory to account not only for varying degrees of hydrodynamic interaction (through the shielding parameter h) but, in addition, to account for departure from Θ-solvent behavior through a parameter ϵ. Tschoegl suggested that, with little loss of accuracy, ϵ could be calculated from

$$\epsilon = (\alpha_\eta^3 - 1)/[(3\alpha_\eta^3 - 2.503) + 2.308(3\alpha_\eta^3 - 2.503)^{1/3}] \tag{4.55}$$

where α_η is the "viscosity expansion factor," defined as

$$\alpha_\eta^3 = [\eta]/[\eta]_\Theta = [\eta]/KM^{1/2} \tag{4.56}$$

It is observed that for many polymer solutions a plot of $[\eta]/M^{1/2}$ vs. $M^{1/2}$ is a straight line, at least over a limited range in M (41). Thus α_η^3 may be written as

$$\alpha_\eta^3 = 1 + bM^{1/2} \tag{4.57}$$

and ϵ may be expressed in terms of M and b as

$$\epsilon = bM^{1/2}/[(3bM^{1/2} + 0.497) + 2.308(3bM^{1/2} + 0.497)^{1/3}] \tag{4.58}$$

As α_η increases from its lower limit of unity ϵ varies from zero to an upper limit of one-third.

Tschoegl and Ferry (42) carried out an experimental test of the modified Zimm theory. Values of ϵ were calculated from intrinsic viscosity data, and values of h were selected which produced curves whose shapes most closely matched the experimental data for η^*. Figure 4.22 shows data obtained with a fractionated polyisobutylene of viscosity average molecular weight 840,000. The results are intermediate between the Rouse and Zimm theories, as expected.

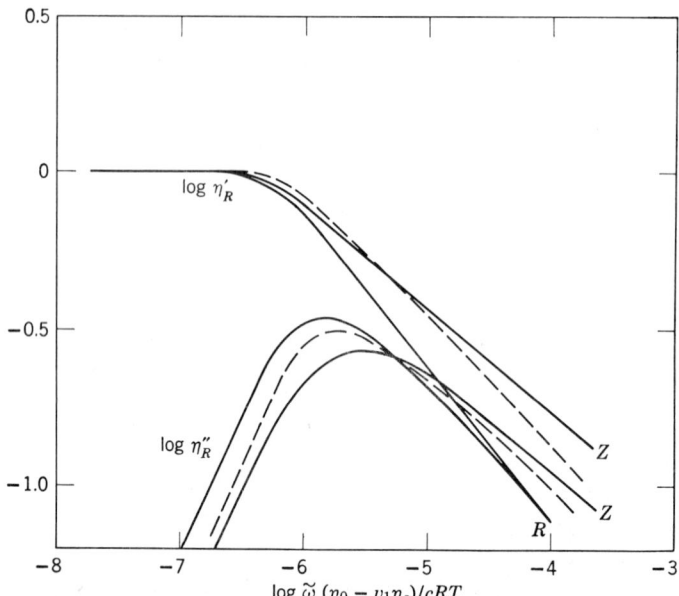

Fig. 4.22. An example (42) of Tschoegl's theory as an intermediate between the Rouse and Zimm theories. The dashed lines are for $h = 5$ and $\epsilon = 0.08$, and are in good agreement with data for 1 and 2% solutions of a fractionated polyisobutylene in a napthenic oil (Primol D).

v_1 is the volume fraction of solvent. The authors use $v_1 \eta_s$ as the solvent contribution to viscosity, instead of simply η_s. The "reduced" components are defined as $\eta'_R = (\eta' - v_1\eta_s)/(\eta_0 - v_1\eta_s)$ and $\eta''_R = \eta''/(\eta_0 - v_1\eta_s)$.

In a more comprehensive experimental study Frederick, Tschoegl, and Ferry (43) measured dynamic moduli of dilute solutions of fractionated polystyrenes over a range of molecular weights, concentrations, and temperatures, and in different solvents, including a Θ-solvent. Figure 4.23 shows a shift from Zimm-like to Rouse-like behavior with increasing molecular weight. Figure 4.24 shows a shift as one goes from a Θ-solvent to a good solvent.

It thus appears that one may incorporate experimental data within the framework of a rational theory, or, equivalently, predict dynamic data with some accuracy, through Tschoegl's extensions of the Zimm theory. As is usual, the price one pays for this is heavy, in that fairly extensive intrinsic viscosity data must be available, including Θ-solvent behavior.

C. Relationship to the Maxwell Model

The relationship of the Rouse-Zimm results to the generalized Maxwell model (Eqs. 3.72 and 3.73) is worth establishing, in view of the still

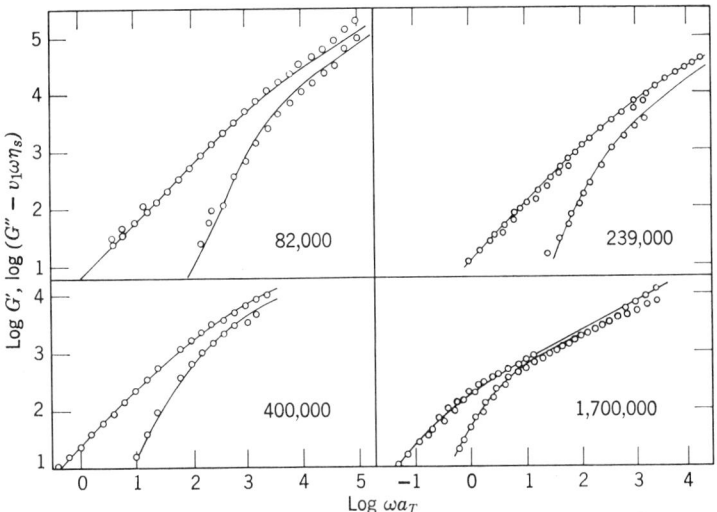

Fig. 4.23. The transition from Zimm- to Rouse-like behavior with increasing molecular weight. Fractionated polystyrenes in Aroclor 1248. All concentrations 2%; all data reduced to 25.0°C. Molecular weights as indicated. Drawn from Ref. 43. Lower curves in each pair are G'.

common usage of the Maxwell model and, in particular, terminology based upon the Maxwell model. If λ_1' in the numerator of Eq. 4.46 is replaced by the right-hand side of Eq. 4.52, the complex viscosity may be written

$$\eta^* - \eta_s = \frac{6(\eta_0 - \eta_s)}{\pi^2} \sum_p \frac{1/p^2}{1 + i\bar{\omega}\lambda_1'/p^2} \quad (4.59)$$

If λ_p' and η_p are now defined by

$$\lambda_p' = \lambda_1'/p^2 \quad (4.60)$$

and

$$\eta_p = 6(\eta_0 - \eta_s)/\pi^2 p^2 \quad (4.61)$$

the complex viscosity may be written as

$$\eta^* - \eta_s = \sum_p \frac{\eta_p}{1 + i\bar{\omega}\lambda_p'} \quad (4.62)$$

If one now looks back at Eq. 3.75 the result above is seen to be identical with it, except for the subtraction of the (usually negligible) term η_s. Hence it is common to interpret Rouse's theory as equivalent to a generalized Maxwell model, with a *discrete* set of relaxation times corresponding to Eq. 4.60.

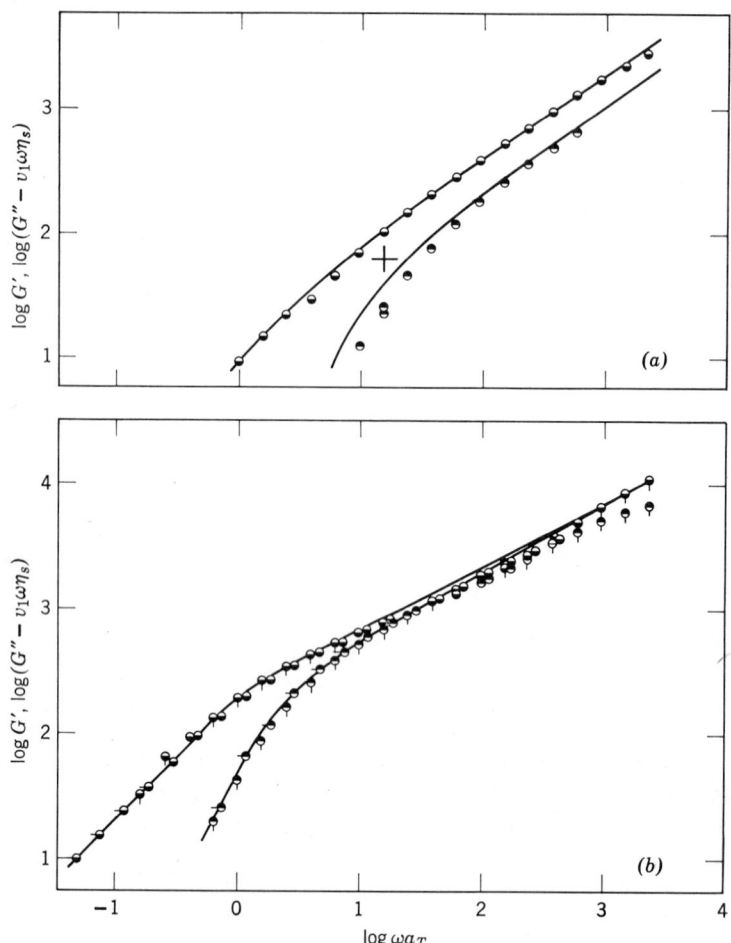

Fig. 4.24. The transition from Zimm- to Rouse-like behavior with increasing solvent power. 2% fractionated polystyrene ($M = 1.7 \times 10^6$) in (a) a Θ-solvent(di-2-ethylhexyl phthalate, DOP, at 12°C.) and in (b) Aroclor 1248. Theoretical curves are drawn with $\epsilon = 0$ and $h = \infty$ in DOP, and $\epsilon = 0.135$ and $h = 0$ in A-1248. Drawn from Ref. 43. Lower curves in each pair are G'.

REFERENCES

1. Flory, P., *Principles of Polymer Chemistry*, Cornell University Press, Ithaca, N. Y., 1953.
2. Billmeyer, F. W., *Textbook of Polymer Chemistry*, Interscience, New York, 1957.
3. Mendelson, R., *Trans. Soc. Rheol.*, **9:1**, 53 (1965).

4. Ballman, R., and R. Simon, *J. Polymer Sci. A*, **2**, 3557 (1964).
5. Einstein, A., *Theory of Brownian Motion*, Dover, New York, 1956, p. 36.
6. Sadron, C., in *Flow Properties of Disperse Systems*, J. J. Hermans, Ed., Interscience, New York, 1953.
7. Frisch, H. L., and R. Simha, in *Rheology*, Vol. I, F. R. Eirich, Ed., Academic Press, New York, 1956, p. 525.
8. Thomas, D., *J. Colloid Sci.*, **20**, 267 (1965).
9. Debye, P., *J. Chem. Phys.*, **14**, 636 (1946).
10. Debye, P., and A. Bueche, *J. Chem. Phys.*, **16**, 573 (1948).
11. Kirkwood, J., and J. Riseman, *J. Chem. Phys.*, **16**, 565 (1948).
12. Hermans, J. J., in *Flow Properties of Disperse Systems*, J. J. Hermans, Ed., Interscience, New York, 1953.
13. Brandrup, J., and E. H. Immergut, Eds., *Polymer Handbook*, Interscience, New York, 1966, Ch. IV.
14. Van Krevelen, D., and P. Hoftyzer, *J. Appl. Polymer Sci.*, **10**, 1331 (1966).
15. Morawetz, H., *Macromolecules in Solution*, Interscience, New York, 1965, p. 308.
16. Bueche, F., *Physical Properties of Polymers*, Interscience, New York, 1962.
17. Bueche, F., *J. Chem. Phys.*, **22**, 1570 (1954).
18. Middleman, S., *J. Appl. Polymer Sci.*, **11**, 417 (1967).
19. Porter, R. S., and J. F. Johnson, *Trans. Soc. Rheol.*, **7**, 241 (1963).
20. Wyman, D., L. Elyash, and W. Frazer, *J. Polymer Sci. A*, **3**, 681 (1965).
21. Dunleavy, J. E., and S. Middleman, *Trans. Soc. Rheol.*, **10:1**, 157 (1966).
22. Brodnyan, J. G., and E. L. Kelley, *Trans. Soc. Rheol.*, **7**, 125 (1963).
23. Brodnyan, J. G., and E. L. Kelley, *Trans. Soc. Rheol.*, **5**, 205 (1961).
24. Porter, R. S., M. J. R. Cantow, and J. F. Johnson, *Proceedings of the Fourth International Congress on Rheology*, Pt. 2, E. H. Lee and A. L. Copley, Eds., Interscience, New York, 1965, p. 479.
25. Robinson, D., M. S. Thesis, University of Rochester, Rochester, N. Y., 1967.
26. Stratton, R. A., *J. Colloid Interface Sci.*, **22**, 517 (1966).
27. Ballman, R., and R. Simon, *J. Polymer Sci. A*, **2**, 3557 (1964).
28. Graessley, W. W., R. L. Hazelton, and L. R. Lindeman, paper presented at Soc. Rheology Meeting, Santa Barbara, Feb., 1967.
29. Graessley, W. W., *J. Chem. Phys.*, **43**, 2696 (1965).
30. Williams, M. C., *A.I.Ch.E. J.*, **12**, 1064 (1966).
31. Fixman, M., *J. Chem. Phys.*, **35**, 889 (1961).
32. Williams, M. C., *A.I.Ch.E. J.*, **13**, 534 (1967).
33. Williams, M. C., *A.I.Ch.E. J.*, **13**, 955 (1967).
34. Huppler, J. D., *Trans. Soc. Rheol.*, **9:2**, 273 (1964).
35. Zimm, B., *J. Chem. Phys.*, **24**, 269 (1956).
36. Rouse, P. E., *J. Chem. Phys.*, **21**, 1272 (1953).
37. Harrison, G., J. Lamb, and A. Matheson, *J. Phys. Chem.*, **68**, 1072 (1964).
38. Tschoegl, N. W., *J. Chem. Phys.*, **39**, 149 (1963).
39. Tschoegl, N. W., *J. Chem. Phys.*, **40**, 473 (1964).
40. Ptitsyn, O., and Y. Eisner, *Zh. Fiz. Khim.*, **32**, 2464 (1958): *Zh. Tekhn. Fiz.*, **29**, 1117 (1959).
41. Stockmayer, W., and M. Fixman, *J. Polymer Sci. C*, **1**, 137 (1963).
42. Tschoegl, N. W., and J. D. Ferry, *J. Phys. Chem.*, **68**, 867 (1964).
43. Frederick, J. E., N. W. Tschoegl, and J. D. Ferry, *J. Phys. Chem.*, **68**, 1974 (1964).

5

Correlation and Interrelation of Material Functions

In previous chapters discussions of various rheological models were accompanied whenever possible by exhibits of experimental data in support of the models. Each of these demonstrations was intended to indicate that the model under consideration did indeed predict material functions bearing some resemblance to observed behavior. In this chapter an attempt will be made to present a more comprehensive view of the rheological behavior of real polymeric fluids. Particular attention will be given to methods of correlation of data.

The complexity and variety of polymer rheology can be anticipated at the outset by reviewing some of the factors upon which the behavior of a polymeric fluid may depend: time scale of the deformation, temperature, degree of polymerization, monomer molecular weight, monomer "size," molecular weight distribution, concentration, solvent power (excluded volume), and solvent viscosity (the last three, if in solution).

If the material is not a linear amorphous polymer, but is capable of developing chemical crosslinks with neighbor molecules, or developing crystalline structure, then additional structural parameters enter to lend variety to the response of such materials. Small wonder then that one may find a specific material which, under certain conditions, behaves in accordance with a particular rheological model! The goal of rheology, however, is not to find any fluid which lends confirmation to a particular theory. Rather it is to provide the means whereby one may describe the mechanical response of some *particular* material of interest subjected to a specific deformation. This is clearly a difficult task, but one to which our attention must now turn.

I. VISCOSITY AT LOW SHEAR RATE

Experimental evidence (1) indicates that the viscosity–molecular weight relationship exhibits two distinct regions as shown in Fig. 5.1. In the low

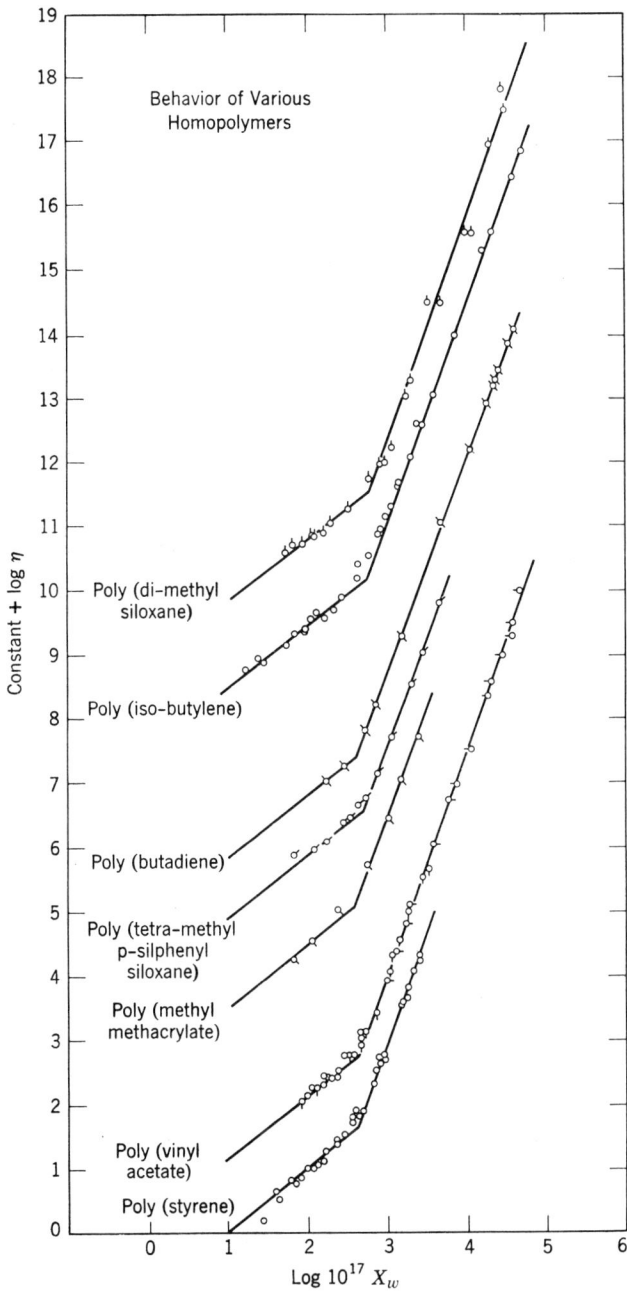

Fig. 5.1. Typical viscosity–molecular weight dependence for molten polymers. X_w is defined in Eq. 5.2, and is proportional to M_w.

molecular weight region, the viscosity increases with a low power of M, generally in the neighborhood of unity. In the high molecular weight region, the viscosity shows a very strong molecular weight dependence, η_0 being proportional to $M^{3.4}$. The transition between these two regions occurs over such a narrow range of molecular weight that one is led to define a "critical" molecular weight M_c as the intersection of the lines representing the high and low molecular weight regimes. The critical molecular weight M_c varies from one polymer to another, being about 4000 for polyethylene and 38,000 for polystyrene (2).

Such a sharp change in the molecular weight dependence must be associated with a corresponding change in the mechanism, on a molecular level, responsible for viscosity. Bueche (3,4) has offered a theory which is in general agreement with observed behavior. The essential idea of his theory is that, above a certain chain length, molecules are sufficiently long to become entangled at various points along their lengths with neighboring molecules. The force required to move an entangled molecule is much greater than that for a shorter, unentangled molecule, because the entangled molecule drags other molecules with it, each of which may itself drag other molecules. The number of primary, secondary, etc., entanglements increases sharply with chain length, and thus the viscosity shows a strong dependence on molecular weight.

The similarity in shape of the viscosity–molecular weight curves for different polymers suggests the possibility that a set of reduced viscosity and molecular weight variables might be found from which a *single* master curve could be presented as a representation applicable to a variety of polymers. Fox and associates (1,5) have made some valuable experimental observations in an attempt to correlate viscosity–molecular weight data. Their results may be written, in agreement with the form of Bueche's theory, as

$$\eta_0/\zeta_0 = 4.8 \times 10^8 (X/X_c)^a \tag{5.1}$$

$$a = 3.4 \quad \text{for} \quad X \geqq X_c$$
$$a = 1.0 \quad \text{for} \quad X < X_c$$

X is defined as

$$X = (s_0)^2 Z \varphi_1 / M v \tag{5.2}$$

where $(s_0)^2$ is the mean square radius of gyration* of an unperturbed molecule, Z is the number of chain atoms in the polymer, v is the specific volume of polymer, and φ_1 is the volume fraction of polymer if the material

---

* $(s_0)^2 = R^2/6$, where R^2 is the mean square end-to-end distance of a polymer molecule (p. 135). Since R^2/M is nearly independent of M (cf. Eq. 4.3), X is nearly proportional to Z, which is proportional to molecular weight.

is in solution. ζ_0 is the friction factor per chain atom.* X_c is the value of X at $Z = Z_c$, and appears to be a universal constant ($X_c = 4.7 \times 10^{-15}$) for a large number of polymers.

The most difficult property to measure *a priori* is the friction coefficient ζ_0. Fox and Allen (1) actually calculate ζ_0 from Eqs. 5.1 and 5.2, and present a correlation for ζ_0 which may be written as

$$\log \zeta_0 = -1.2E/2.3RT - 11.4\{\alpha_f(T - T_g)/[0.024 + \alpha_f(T - T_g)]\} \quad (5.3)$$

In order to calculate ζ_0 it is necessary to know the activation energy E, the glass temperature† T_g, and the expansion coefficient α_f.‡

The limitations to the use of Eq. 5.3 are not yet established. In the absence of any information one may take a value for E of 4.4 kcal/mole and expect an error in ζ_0 of as much as a factor of 4 (1). Values of α_f for many polymers lie in the range $3-9 \times 10^{-4}$ cm.³/g.-°K. Simha and Boyer (7) find a correlation between α_f and T_g, and suggest $\alpha_f T_g = 0.113$ for the prediction of α_f from a knowledge of the glass transition temperature. Table 5.1 gives values of T_g for several polymers.

The dependence of η_0 on volume fraction of polymer in the case of diluted polymer is given partially by Eqs. 5.1 and 5.2. If ζ_0 were independent of φ_1, then $\eta_0 \sim \varphi_1^a \sim c^a$, and η_0 would increase with the 3.4 power of φ_1 for $X > X_c$. For high concentrations, however, one commonly observes $\eta_0 \sim c^5$, due to a significant dependence of ζ_0 itself on c. ζ_0 also depends on the nature of the solvent. A useful correlation of solution viscosities at low shear rates is given by

$$\eta_r = \eta_0/\eta_s = f(cM^\beta) \quad (5.4)$$

β is usually given as 0.68 (8), but more recent studies (9) indicate that β

---
* In terms of the monomeric friction factor $\zeta_0 = \zeta/n_a$, where n_a is the number of chain atoms per monomer.

† If a liquid is cooled continuously and if crystallization does not occur, it solidifies to a glasslike state over a fairly narrow temperature range characteristic of the material. The temperature T_g about which this takes place is defined as the glass temperature (sometimes called the second-order transition temperature). T_g may vary by a few degrees, depending on the cooling rate, and so is not precisely defined. Properties such as the thermal expansion coefficient (the slope of the specific volume *vs.* temperature curve) or the heat capacity exhibit abrupt changes at T_g by which this temperature may be established.

‡ A major factor affecting ζ_0 is the amount of "free volume" in which a polymer chain, or a segment of a chain, may move. Free volume is not directly measurable. It is believed, however, to be related to the thermal expansion coefficient of the material. Thus, while α_f may be regarded as an empirical constant, it is generally calculated as the difference, $\alpha_l - \alpha_g$, between the thermal expansion coefficients of the liquid and the glass. A discussion of the glass transition and free volume is given by Ferry (6, p. 218).

may vary from one material to another in the range 0.54 to 0.72.* Figure 5.2 shows solution viscosity data for polyisobutylene (10) plotted according to Eq. 5.4. Similar data exist for solutions of polystyrene (11), polyvinyl

Table 5.1
Glass Transition Temperatures

Polymer	T_g (°K.)
Polyethylene	148
Polydimethylsiloxane	150
Polybutadiene	178
Polyisobutylene	200
Polyethylene oxide	206
Polyvinyl acetate	301
Polyvinyl chloride	354
Polystyrene	373
Polymethyl methacrylate	378

Values given here are typical. There is only a slight molecular weight dependence at the molecular weights of usual interest.

acetate (12), and acrylonitrile–methyl methacrylate copolymers (13). At large values of $cM^{0.68}$ it is usually found that $\eta_r \sim (cM^{0.68})^5 = c^5 M^{3.4}$. A particularly interesting set of data is that of Longworth and Busse (14) for polyethylene in paraffin wax. The concentration varies from dilute solution to melt. The high density and low density materials are quite different, as shown in Fig. 5.3.

Insufficient evidence exists at the present time to allow the construction of a master curve for η_r which would be applicable to a number of different polymers. Ferry and coworkers (11) suggested in this regard that one might plot solution viscosity data for different polymers as η_r vs. $c^5 Z^{3.4}$. The use of the chain length in place of molecular weight results in a good correlation of data for a number of different polymers in the region $\eta_r < 10^4$. In the high molecular weight and high concentration region, however, data for various polymers, and even data for the same polymer in different solvents, show relative viscosity differences greater than an order of magnitude. Until more experimental data are available, particularly for highly concentrated solutions, it is unlikely that a useful generalized correlation of solution viscosity will be constructed.

It is interesting to note that data plotted as in Fig. 5.2 do not show clearly a critical value of $cM^{0.68}$ at which "entanglement" occurs. This

* A molecular theory due to Williams (15) predicts $\eta_0 = f(cM^{0.625})$ (see Chapter 4, p. 159).

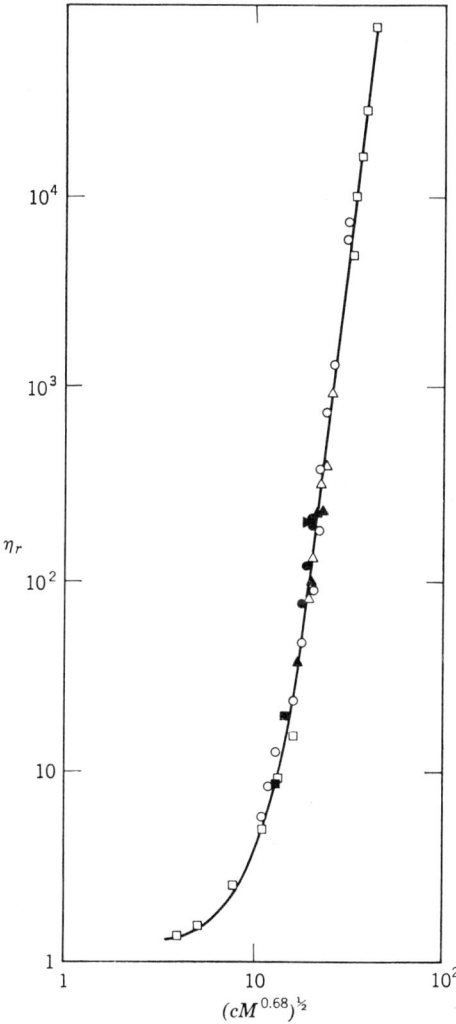

Fig. 5.2. Relative viscosities (at zero shear) of solutions of polyisobutylene. Molecular weights (viscosity average) are (○) 7.3×10^5 ($L100$), (△) 1.25×10^6 ($B100$), and (□) 5.1×10^6 ($B200$). Open symbols are toluene solutions, filled symbols are tetralin solutions.

may be a real effect due to smoothing of this transition in solutions. It seems more likely that the apparent sharp transition shown in $\eta_0 - M$ curves is really an artifact of the data arising from the small number of data points usually available in the transition region. A very complete study of the zero shear viscosities of molten polyethylene oxides in the "entanglement" region shows smooth curvature rather than a sharp break (16). A recent study of molten polybutadiene also fails to reveal a

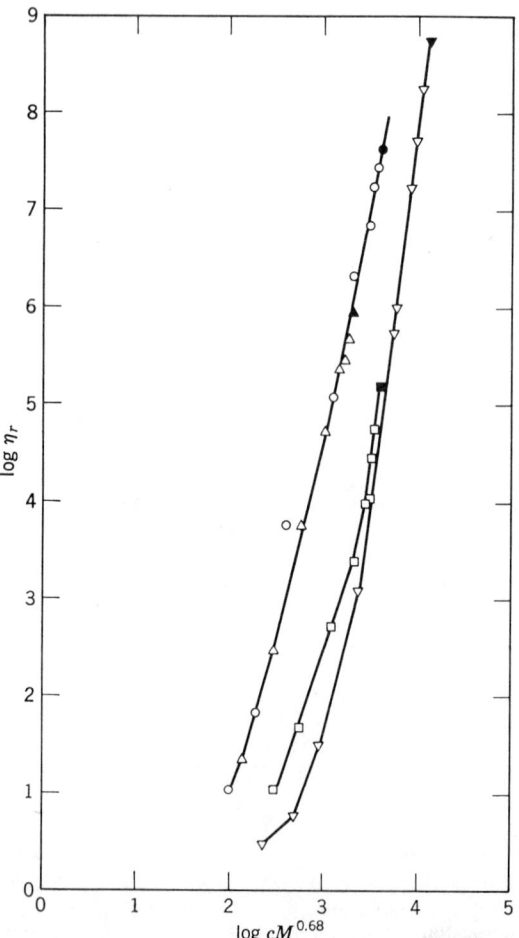

Fig. 5.3. Viscosities of polyethylene–paraffin wax solutions (14). The solid symbol represents the molten undiluted polymer. High density: (○) $M_w = 1.9 \times 10^5$; (△) $M_w = 7 \times 10^4$. Low density: (▽) $M_w = 10^6$; (□) $M_w = 2 \times 10^5$.

sharp transition (17). However, this uncertainty does not reduce the value of M_c as a parameter of some value in the correlation and prediction of experimental results, at least in the regions removed from the entanglement region.

A. Dependence on Temperature

Viscosity is a very strong function of temperature, and the ability to predict the variation of viscosity with some expected temperature variation

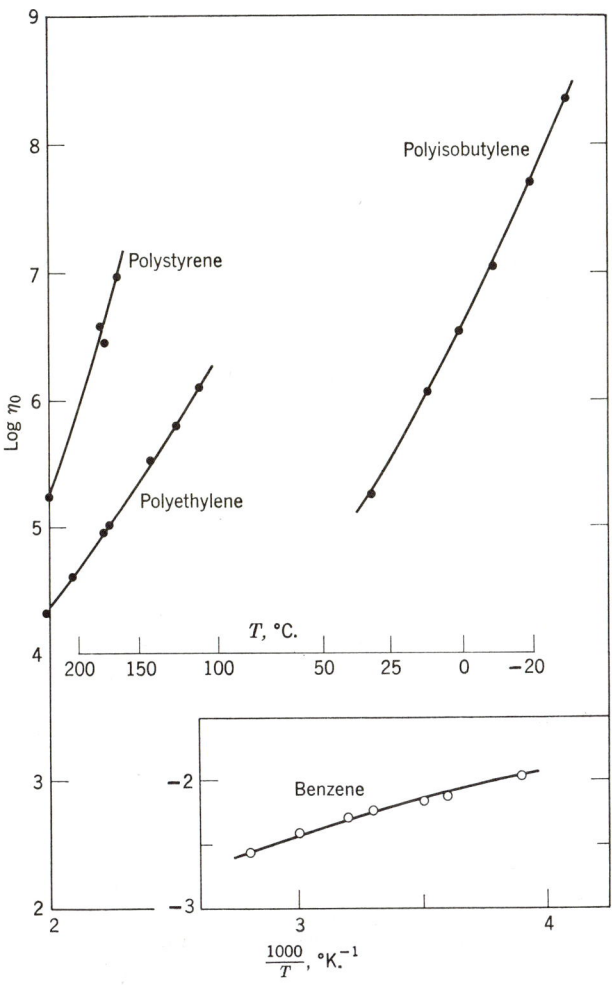

Fig. 5.4. Temperature dependence of viscosity.

is both critical and difficult. Simple organic liquids show a temperature dependent viscosity such as that illustrated in Fig. 5.4 for benzene. Over a temperature range of the order of 50°C. one can represent such behavior by an expression of the form (18)

$$\eta(T) = A \exp(B/T) \quad (5.5)$$

The viscosity at some temperature relative to its value at some reference temperature T_r would then be

$$\eta(T)/\eta(T_r) = \exp[-B(T - T_r)/TT_r] \quad (5.6)$$

For simple low molecular weight liquids, the constant B may be estimated from thermodynamic principles (18).

High polymers show a similar temperature dependence of viscosity, as shown in Fig. 5.4 for typical polymers. The coefficient B depends upon structural and thermal properties of the polymer, and cannot be predicted *a priori* from basic thermodynamic considerations. For this reason other mathematical representations of viscosity are used, in which coefficients appear that can be related to, *and* predicted from, structural considerations.

An important study of the temperature-dependent properties of polymers is the work of Williams, Landel, and Ferry (19). They were particularly interested in the temperature dependence of the relaxation time associated with viscoelastic response, and measured a coefficient a_T defined by

$$a_T = \eta T_r \rho_r / \eta_r T \rho \quad (5.7)$$

where the subscript r means that the quantity to which it is appended is measured at the temperature T_r. a_T is essentially the shift factor defined in Chapter 4 (Eq. 4.40), and represents the ratio of relaxation times at temperatures T and T_r.*

Because of the relatively small temperature dependence of $T_r \rho_r / T \rho$ (T is in °K.), a_T is very nearly the same as $\eta(T)/\eta(T_r)$ and should be given by a function similar to Eq. 5.6. The significant feature of the work reported by Williams and coworkers was the discovery that, if a special choice was made for the reference temperature T_r, the function a_T was a *universal* function of $T - T_r$ obeyed by a variety of polymeric materials, as well as by several inorganic glasses.

The value of T_r was determined for each material by trial and error, using data for a high molecular weight polyisobutylene as a standard, with $T_r = 243°K$. The essential point is that values of T_r could be found for

---

* The solvent viscosity η_s which appears in Eq. 4.40 is dropped in the case of undiluted polymer, and is commonly neglected in comparison to η_0 even in the case of solutions. The temperature variation of ρ/ρ_r is usually neglected also.

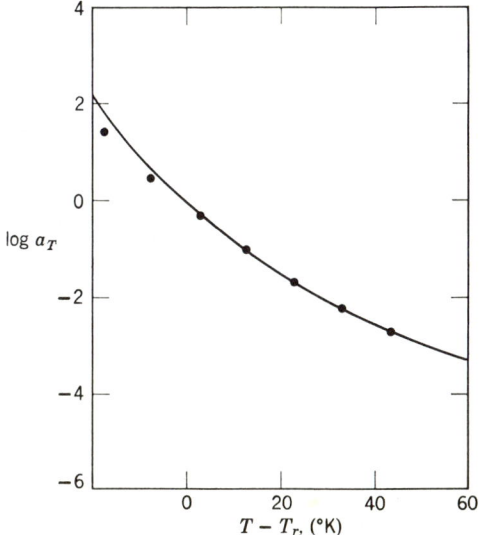

Fig. 5.5. The WLF equation. Solid symbols represent data of Leaderman (47) for polyisobutylene.

many materials which resulted in a_T vs. $T - T_r$ curves nearly identical with the reference curve for the polyisobutylene. The second significant observation in this study was that, within a standard deviation of $\pm 5°$K., the reference temperature was related to the glass temperature T_g by $T_r = T_g + 50$.

Figure 5.5 shows a_T vs. $T - T_r$. It should be noted that an uncertainty in T_r of $\pm 5°$K. corresponds to an uncertainty in a_T of as much as half an order of magnitude (a factor of 3). Thus, while Fig. 5.5 offers a powerful tool for the estimation of the temperature effect on viscosity, the resultant estimate may be in considerable error. To place this point in its proper perspective, however, one must note that Fig. 5.5 covers a range of viscosities of the order of 10^5. Thus a factor of 3 is seen to be small in comparison to a factor of 10^5.

The equation recommended for a_T is known as the WLF equation, and is written as

$$\log a_T = -c_1(T - T_r)/(c_2 + T - T_r) \qquad (5.8)$$

with $c_1 = 8.86$ and $c_2 = 101.6$. Equation 5.8 should not be used below $T = T_g$ nor above $T = T_g + 100$. The curve in Fig. 5.5 is drawn using these particular values of c_1 and c_2. For comparison, experimental data of Leaderman et al. (47) are shown to indicate a case for which the data are considered to be in good agreement with the empirical "universal correlation."

If one wishes to make the approximation that $T_r = T_g + 50$, then the WLF equation becomes

$$\log a_T = -17.4\,(T - T_g)/(51.6 + T - T_g) \qquad (5.9)$$

This is less accurate than Eq. 5.8 because of the inaccuracies inherent in the measurement of T_g, and (more importantly) inaccuracy in the approximation $T_r = T_g + 50$. It does, however, provide a means of estimating the temperature dependence of the viscosity of a material for which T_g is known, but for which so little information is available that T_r is not available. Table 5.2 gives some values of T_r.

Table 5.2
Reference Temperature for the WLF Equation (19)

Polymer	T_r (°K.)	$T_r - T_g$ (°K.)
Polyisobutylene		
$\overline{M}_v = 4170$	238	44
$\overline{M}_v = 8500$	240	38
Polystyrene	407	45
Polymethyl methacrylate	433	55
Polyvinyl acetate	349	48
Butadiene–styrene copolymer (75/25)	268	57

The values of T_g used here do not necessarily correspond to those of Table 5.1.

Equation 5.9 bears some resemblance to Eq. 5.3. The latter equation is more accurate than the WLF equation, but requires a knowledge of the parameters E and α_f. For this reason the WLF equation is more commonly used for estimating the temperature effect on viscosity. For temperatures above $T_g + 100$ an equation of Fox and Loshaek (20) is recommended:

$$\log a_T = C - B/T^{1+\alpha} \qquad (5.10)$$

The constant α is usually observed to be in the neighborhood of one to two although larger values are sometimes reported (20, p. 451).

II. NON-NEWTONIAN VISCOSITY

Above some critical shear rate, polymeric materials depart from newtonian behavior and exhibit a viscosity which decreases with increasing shear rate. Most molecular theories, as discussed in Chapter 4, imply

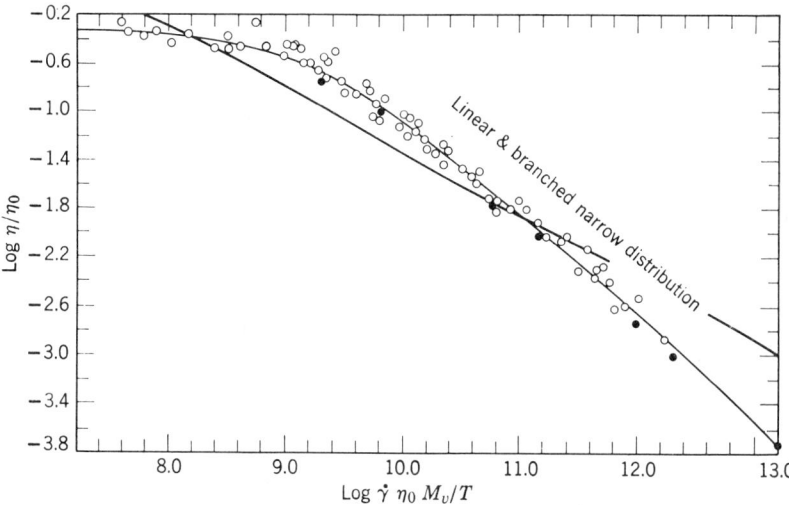

Fig. 5.6. Superposition of data for monodisperse polystyrenes. The heavier line is the Bueche theory. Adapted from Fig. 2 of Ref. 21.

that the viscosity–shear rate curve is a universal function when plotted as η/η_0 vs. $\dot{\gamma}\lambda_1$, where λ_1 is a characteristic relaxation time for the fluid. Bueche's theory, for example, gives λ_1 as proportional to $\eta_0 M/\rho RT$ for a molten polymer (Eq. 4.39).

A number of tests have been made of the effectiveness of $\dot{\gamma}\lambda_1$ as a superposition variable for non-newtonian viscosity. Figure 5.6 shows data of Wyman, Elyash, and Frazer (21) and of Ballman and Simon (22), obtained with monodisperse polystyrenes. The use of $\dot{\gamma}\eta_0 M_v/T$ as a "reduced" shear rate leads to a correlation of the available data.* Agreement with the Bueche theory is not bad; the data cross the Bueche curve at various points on the shear rate axis. It is interesting to note that the "broad distribution" polymer (Dow Styron-666) studied by Wyman and coworkers gave a viscosity–shear rate curve which was in excellent agreement with the Bueche theory. This is a surprising result, in view of the fact that the Bueche theory is derived by assuming a monodisperse polymer.

Gruver and Kraus (23) present data for linear monodisperse polybutadienes of molecular weights in the range 52,000 to 740,000. By taking the data as η/η_0 vs. $\dot{\gamma}$, and shifting the data along the $\dot{\gamma}$-axis until coincidence with the Bueche theory was obtained for each fluid, an estimate of

* A better correlation is obtained, however, using $\dot{\gamma}\eta_0 M_v^{0.75}/T$, as noted in Chapter 4, p. 156.

λ_1 (which would just be the shifting factor) could be obtained. λ_1 was found to increase with $(M\eta_0)^{1/2}$, rather than linearly with $M\eta_0$, as demanded by the Bueche theory. Furthermore, λ_1 was found to be one to two orders of magnitude below the value which would be predicted by the Bueche theory. This means that the data, if plotted as η/η_0 vs. $\dot{\gamma}\lambda_1$, with λ_1 calculated from Eq. 4.39, would lie above and to the right of the Bueche curve, shifted along the $\dot{\gamma}\lambda_1$-axis by one to two orders of magnitude. Although much more pronounced, this is the same type of behavior that one will observe with other materials. Similar large discrepancies are observed in a study of monodisperse polyisoprene (24).

By way of contrast a study (17) of *low* molecular weight polybutadienes (M_v in the range 5000 to 8000) shows close conformity to the Bueche theory, with departures from newtonian flow occurring in the shear rate region predicted by theory. These materials were not monodisperse, although they were of fairly narrow distributions, having M_w/M_n of the order of 1.25. Data taken with a polybutadiene of a broader distribution ($M_w/M_n \sim 2$) showed an earlier departure from newtonian flow and a more shallow viscosity drop-off with shear rate, in accordance with theoretical predictions of Chapter 4 (p. 154).

The viscous flow of molten polypropylene was studied by Vinogradov and Prozorovskaya (25), using a sample with a viscosity average molecular weight of 4.5×10^5, at a number of temperatures and over a range of shear rates. Their data are shown in Fig. 5.7, and indicate a good internal consistency with respect to correlation with $\dot{\gamma}\lambda_1$. Agreement with the Bueche theory is excellent.

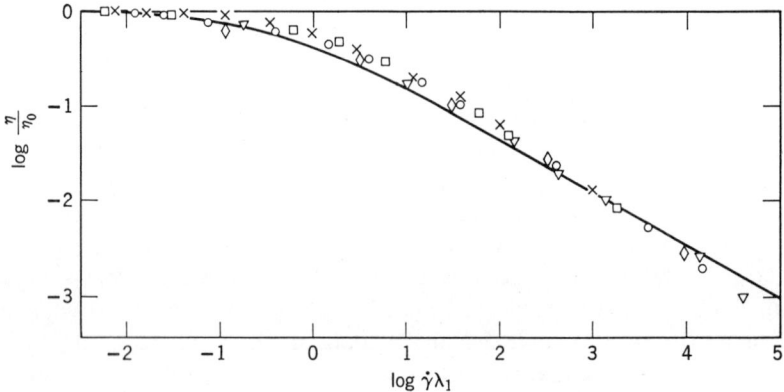

Fig. 5.7. Superposition of data for molten polypropylene. Curve is Bueche theory. Temperatures (°C.) are: (×) 270, (□) 250, (○) 230, (◊) 210, and (▽) 190.

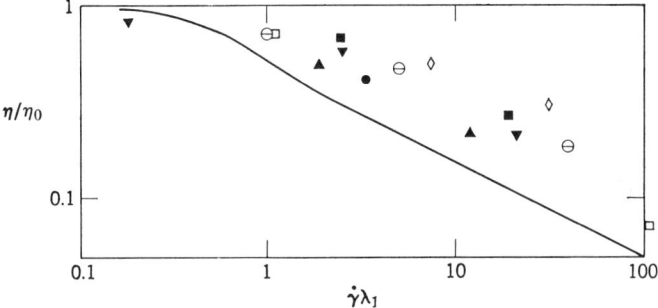

Fig. 5.8. Superposition of data for molten polydimethylsiloxane (26; ●, ▲, ▼, ■) and polyethylene (28; ◊, ⊖, □). Curve is Bueche theory.

Ito (26) presented flow curves for molten polydimethylsiloxane over a wide range of molecular weights. Figure 5.8 shows some of these data plotted in terms of reduced variables. Again one sees the data above the Bueche theory.

Molten polyethylene, studied by Smith and McKelvey (27, p. 36), shows good quantitative agreement with the Bueche theory. Other polyethylene data, presented by Mendelson (28), exhibit behavior like that of polydimethylsiloxane, as shown in Fig. 5.8.

In addition to studies of *molten* polymers, considerable attention has also been directed toward the viscosity of polymer solutions. The superposition of dilute solution data for polyisobutylenes of various molecular weights is treated in Fig. 5.9 (10). The samples studied were of different broad molecular weight distributions, which undoubtedly accounts for the failure to observe a tight correlation.

Concentration affects λ_1 primarily through the concentration dependence of η_0, which, in the concentration range usually studied, is of the order of c^3 to c^5. The Bueche theory for solutions gives, in addition, a linear term in c appearing in the denominator of λ_1. This appears to be open to question, there being some evidence that at high concentrations a c^2 dependence is more appropriate* (29). This implies that one must be cautious in using dilute solution data, correlated in terms of reduced variables, to predict the behavior of the melt, since such a long extrapolation with respect to concentration would be required.

Figure 5.10 shows a reduced variable correlation based on data presented by Seely (30) for dilute solutions of polybutadiene. The Bueche theory correlates the data quite well up to moderate shear rates, whereupon

---

* This point is in agreement with a theory of Williams (15) (see p. 160).

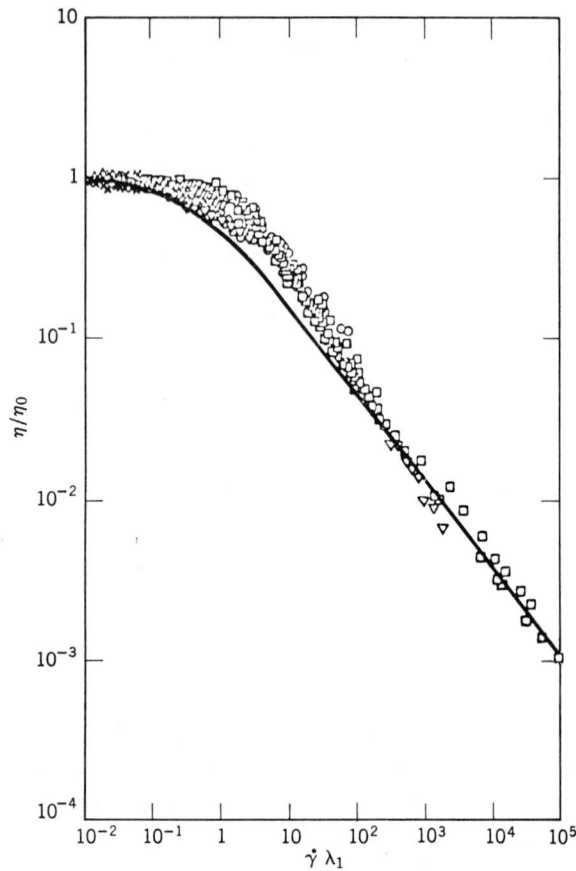

Fig. 5.9. Superposition of data for solutions of various polyisobutylenes in toluene (10). Molecular weights range from 9×10^4 to 5.1×10^6. Curve is Bueche theory.

the theory predicts behavior more non-newtonian than is observed. The scale of the reduced variable plot hides the fact, noted by Seely, that the concentration dependence of the low molecular weight data is not in agreement with that predicted by the Bueche theory. In this case the data behave in accordance with a relaxation time which has no explicit dependence on concentration, $\lambda_1 \sim \eta_0 M_v/T$, rather than $\lambda_1 \sim \eta_0 M_v/cT$. Seely's high molecular weight data, on the other hand, do correlate using $\lambda_1 \sim \eta_0 M_v/cT$.

Brodnyan and Kelley (31) present data for a water-soluble polyelectrolyte, polyacrylic acid, at various degrees of neutralization in sodium hydroxide. They note a slight concentration dependence of the deviation

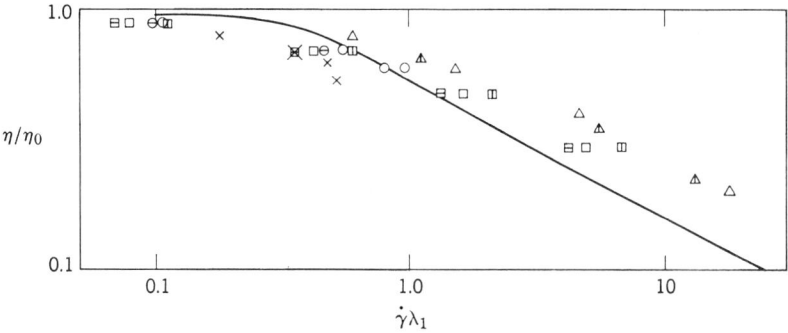

Fig. 5.10. Superposition of data for solutions of polybutadiene (30). Curve is Bueche theory.

of the data from the theory which would be consistent with the use of a power of c less than unity in the denominator of λ_1.

Longworth and Busse (14) present a study of polyethylene diluted with a paraffin wax over a very wide range of concentrations. Their viscosity–shear rate data correlate reasonably well, but lie below and to the left of the Bueche theory, showing departure from newtonian flow at a shear rate an order of magnitude lower than that predicted. This could be due to the method used by the authors to calculate the weight average molecular weight. The solvent (paraffin wax with a molecular weight of 435) was not treated as a diluent; the molecular weight of the mixture was taken as a weight average of polymer and solvent. Hence, the molecular weights used were too low, and the reduced variable correlation should be shifted somewhat to the right. In falling *below* the Bueche theory the data are at variance with the *molten* polyethylene data of Mendelson presented in Fig. 5.8.

What conclusions can be drawn from these demonstrations? In an act of rheological cowardice one could claim the need for more definitive data, especially under conditions of established molecular weight distributions and, for solutions, over the entire range of concentration for a given polymer. On the basis of the data available, however, it is clear that a reduced variable plot, using η/η_0 vs. $\dot{\gamma}\lambda_1$, with $\lambda_1 \sim \eta_0 M_v/cT$ (and c replaced by ρ for melts), does correlate data obtained over a wide range of viscosity, for a given sample of polymer. Two polymers of the same name (such as "polyethylene") may differ in some subtle structural feature*

---

* Differences in polymerization conditions or type of catalyst employed may affect polydispersity and degree of branching.

and show deviations, along the $\dot{\gamma}\lambda_1$-axis, of an order of magnitude. In the absence of any data the Bueche theory gives a reasonable quantitative estimate of the viscosity–shear rate curve for polymer melts, sometimes lying above and sometimes below the experimental curve. Polymer solutions show a generally similar agreement with the Bueche theory, although there is some question in regard to the manner in which concentration enters the relaxation time λ_1 used to form the reduced shear rate $\dot{\gamma}\lambda_1$.

III. NORMAL STRESS COEFFICIENTS

A. Behavior in the Limit of Zero Shear Rate

While normal stress phenomena have been recognized for quite some time, it is only in recent years that reliable experimental techniques and associated dynamic analyses have been developed to the point that significant experimental data may now be obtained. Relatively few normal stress studies have been performed with materials sufficiently well defined so that it is possible to attempt any correlation of the data in terms of parameters such as molecular weight and concentration. Most theoretical analyses (reviewed in Chapters 3 and 4) agree that the zero shear normal stress coefficient $\psi_{12}^{(0)}$ should be given by $c_2\eta_0\lambda$, where λ is the dominant relaxation time of the fluid and c_2 is a constant of the order of unity. The open question, of course, is the manner in which λ is related to measurable molecular parameters.

If the theory of Bueche is taken for the sake of example, and λ calculated from $\lambda_1 = 12\eta_0 M/\pi^2 cRT$ (Eq. 4.39), then a *testable* prediction is (taking $c_2 = 1$)

$$\psi_{12}^{(0)} = 12\eta_0^2 M/\pi^2 cRT \tag{5.11}$$

Figure 5.11 shows a test of this prediction for data covering six orders of magnitude in $\psi_{12}^{(0)}$. The failure of Bueche's relaxation time is evident, yet the general trends exhibited by these data clearly support the prediction that $\psi_{12}^{(0)} \sim \eta_0\lambda$, with λ proportional to, but not necessarily equal to, Bueche's λ_1.

It may be recalled (Chapter 4, p. 161) that the Rouse–Zimm theory gives a λ proportional to λ_1, and that the proportionality constant depends upon the degree of hydrodynamic interaction (Tschoegl's theory, p. 163) to which the polymer molecule is subjected by its solvent. If accounted for, these effects would lead to predicted values of $\psi_{12}^{(0)}$ larger than those predicted by Bueche's λ_1, consistent with the trends observed in Fig. 5.11. However, the order of magnitude of the deviations seen in Fig. 5.11 cannot be explained entirely on this basis.

Much of the observed scatter of these data can probably be attributed to

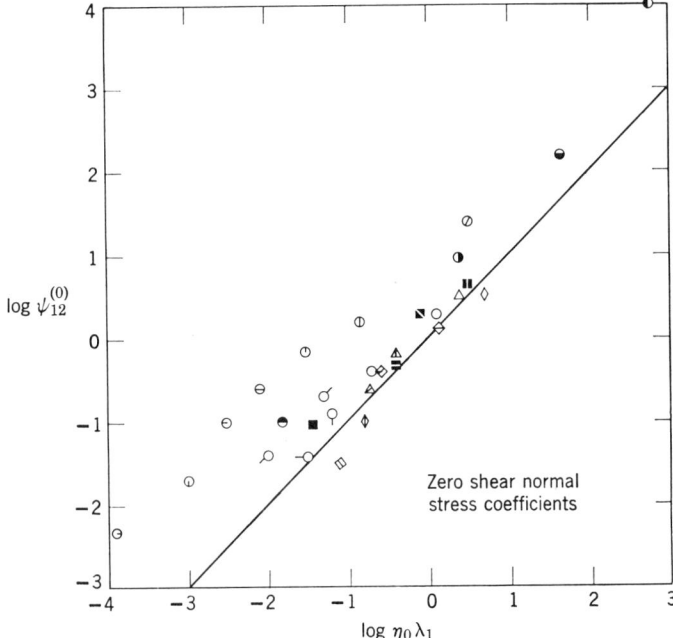

Fig. 5.11. Zero shear normal stress coefficient ψ_{12}^0 for polymer solutions. Polyisobutylene: in cetane (36), (■, ▬, ∎), $M_v = 1.2 \times 10^6$; in decalin (34), (◼) $M_v = 7 \times 10^5$. Polystyrene: in Aroclor 1248 (39), (△, ▲, ◮) $M_v = 1.2 \times 10^6$, (◊) $M_v = 5 \times 10^6$, (♦) $M_v = 5.4 \times 10^5$, (♦) $M_v = 2.4 \times 10^5$, (♦) $M_v = 6.6 \times 10^5$, (♦) $M_v = 2.7 \times 10^5$; in dioctyl phthalate (32), (○, ○, ○, ○) $M_v = 5 \times 10^6$, (○) $M_v = 1.2 \times 10^6$, (○) $M_v = 6.6 \times 10^5$; in toluene (37), (⊖, ⊙, ⊖, ⊙) $M_v = 2.3 \times 10^5$, (●, ◐, ◓, ◑) $M_v = 1.8 \times 10^6$; in decalin (42), (⊖, ⊕, ⊘) $M_v = 2.2 \times 10^5$.

heterogeneity of the molecular weight distributions of the polymers studied. Tamura (32) points out that even a moderate degree of heterogeneity ($M_v/M_n \sim 2$) can lead to an increase in the expected value of $\psi_{12}^{(0)}$ of as much as a factor of 5. None of the materials studied, however, has a well-defined molecular weight distribution, and in this sense these data do not really constitute a critical test of any molecular theory. Despite this reservation it seems worthwhile to point out certain observations, if not conclusions, which may be drawn from the data shown in Fig. 5.11.

The data which show the most consistent departures from the bulk of the data shown are those for highly concentrated (20–35%) solutions of polystyrene. The value of $M_v/M_n = 1.3$ is much smaller than that for some of the other data shown, so it seems unlikely that the differences can be attributed simply to heterogeneity. Dilute solutions in highly viscous solvents show good correlation with theoretical expectations. Dilute polyisobutylene solutions show good agreement with dilute polystyrene solutions. This leads to the conjecture that the concentration dependence of normal stresses is uncertain in highly concentrated solutions, as seems to be the case as well with the non-newtonian viscosity of concentrated solutions.

Vinogradov (25) presents normal stress data for molten polypropylene, obtained by measuring the expansion of the extrudate from a capillary. The normal stress coefficient is many orders of magnitude smaller than would be predicted from $\psi_{12}^{(0)} = \eta_0 \lambda_1$. Abnormally low values of normal stress have also been reported for jets of molten polyethylene (33). It seems likely that jet expansion experiments are not reliable when performed with *molten* polymers. The major problem would seem to be due to the rapid "freezing" of the extrudate, which may retard the complete expansion of the jet. Furthermore the extreme viscosity of melts makes it impossible to achieve very high speed jets. The slowly extruded polymer falls vertically from the capillary exit in a frozen "rope," and may exert considerable pull on the molten polymer near the exit, further reducing the degree of expansion of the jet. Unfortunately, no normal stress measurements with molten polymers appear to have been performed in reliable instruments such as the cone and plate or parallel plate rheometers.

B. Dependence on Shear Rate

In accordance with a number of constitutive equations, both continuum (e.g., Eq. 3.82) and molecular (Graessley, p. 156, and Williams, p. 158), the normal stress coefficient ψ_{12} should decrease with increasing shear rate. Furthermore, the rate of decrease of ψ_{12} should be about twice as great as the rate of decrease of η with shear rate. Few data are available from which $\psi_{12}/\psi_{12}^{(0)}$ may be calculated as a function of shear rate. Figure 5.12 shows data of Ginn and Metzner (34), Shertzer and Metzner (35), and Markovitz and Brown (36), replotted as $\psi_{12}/\psi_{12}^{(0)}$ vs. $\lambda_1 \dot{\gamma}$. The curves clearly show the more rapid drop-off of the normal stresses as compared to the viscosity decrease.

Markovitz and Brown (36) investigated the concentration superposition of normal stress data obtained with cetane solutions of polyisobutylene. They reduce the stress data directly, rather than calculating zero shear normal stress coefficients and using $\psi_{12}/\psi_{12}^{(0)}$. Their procedure corresponds in principle, however, to that used in preparing Fig. 5.12. Their results indicate an excellent correlation of the data, but require a c^2 term in the denominator of λ_1, rather than a linear term. Their treatment of the normal stress data of Kotaka (37) requires a c^3 term, instead of c, in order to obtain adequate superposition of the data. Kotaka's data were obtained

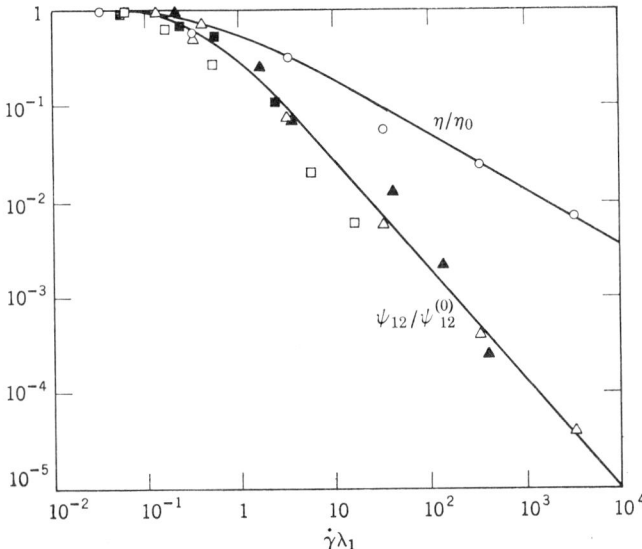

Fig. 5.12. Shear rate dependence of ψ_{12} for various polyisobutylene solutions. The lower curve is almost exactly equal to $(\eta/\eta_0)^2$. PIB in decalin, $M_v = 7 \times 10^5$: ($\bigcirc$, $\triangle$) 5% (34), ($\blacktriangle$) 3% (35). PIB in cetane, $M_v = 1.2 \times 10^6$: (36) ($\square$) 6.9%, ($\blacksquare$) 5.4%.

with concentrated polystyrene solutions, and are identical with the data of Fig. 5.11 which show the greatest departure from the Bueche theory. Hence their results are just another reflection of the uncertainty of the concentration dependence of normal stress coefficients.

IV. RELATIONSHIP OF STEADY SHEAR TO DYNAMIC PROPERTIES

The viscosity η and normal stress coefficients ψ_{ij} are steady shear properties, in the sense that they are well defined only in the case of a steady simple shearing flow. In Chapter 2 it was noted that dynamic tests, in which a sinusoidally varying simple shear flow is established, are also used to investigate the properties of polymeric materials. In Chapter 3 it was seen that certain continuum theories predict relationships among the steady and dynamic coefficients. The degree to which these relationships exist will be examined here.

Dynamic coefficients are usually expressed in terms of the complex dynamic viscosity $\eta^* = \eta' - i\eta''$, or the complex dynamic (shear) modulus $G^* = i\tilde{\omega}\eta^* = G' + iG''$. The real part G' is called the storage modulus, and is related to the imaginary part of the dynamic viscosity by $G' = \tilde{\omega}\eta''$.

The imaginary part G'' is called the loss modulus, and is given in terms of η' by $G'' = \bar{\omega}\eta'$. The expected qualitative behavior of η^* and $\mathbf{G}^*$ is illustrated in Chapter 4, Figs. 4.20 and 4.21.

The general Coleman-Noll theory (p. 91) predicts the following relationships between steady shear and dynamic coefficients:

$$\eta_0 = \lim_{\bar{\omega} \to 0} G''/\bar{\omega} = \lim_{\bar{\omega} \to 0} \eta' \tag{5.12}$$

$$\psi_{12}^{(0)} = \lim_{\bar{\omega} \to 0} 2G'/\bar{\omega}^2 = \lim_{\bar{\omega} \to 0} 2\eta''/\bar{\omega} \tag{5.13}$$

Stronger predictions are made by the continuum theory of Spriggs (p. 104),

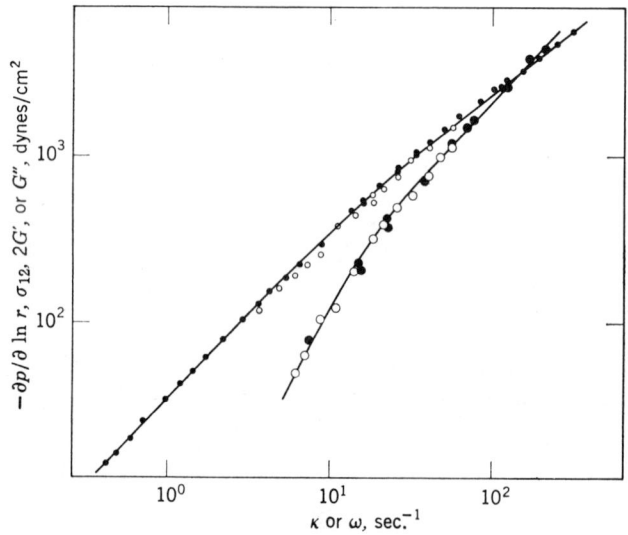

Fig. 5.13. Test of Eqs. 5.14 and 5.15, for a solution of polybutadiene in xylene. (●) and (○) compare shear stress (σ_{12}) and loss modulus (G'') as functions of shear rate (κ) or frequency (ω). (●) and (○) compare normal stress ($-\partial p/\partial \ln r$, see Eqs. 2.120–2.122) and storage modulus ($2G'$). Drawn from Ref. 38.

which suggests that

$$\eta(\dot{\gamma}) = \eta'(c\bar{\omega}) = G''(c\bar{\omega})/c\bar{\omega} \tag{5.14}$$

$$\psi_{12}(\dot{\gamma}) = 2\eta''(c\bar{\omega})/c\bar{\omega} = 2G'(c\bar{\omega})/(c\bar{\omega})^2 \tag{5.15}$$

without any restriction to vanishingly small values of $\dot{\gamma}$ or $\bar{\omega}$. c is a "shift factor," and Eq. 5.14 states that η and η' have the same shape but are shifted with respect to each other, along the $\dot{\gamma}$-$\bar{\omega}$ axis, by the factor c.

Figure 5.13 shows a comparison between steady shear and dynamic shear behavior of a solution of polybutadiene (38). The agreement is excellent for both sets of functions well into the shear dependent regions. No shift factor is required. Such behavior is consistent with the Weissenberg hypothesis ($\psi_{23} = 0$) and Eqs. 5.14 and 5.15.

A similar comparison of data for a polymethyl methacrylate solution (38) is shown in Fig. 5.14. Here the curves cannot be shifted into coincidence, and Eqs. 5.14 and 5.15 are clearly refuted. The trend of the date, as $\dot{\gamma}$ and $\bar{\omega}$ decrease, is toward agreement with Eqs. 5.12 and 5.13, but insufficient low shear data are available for a firm confirmation.

There seems to be little question that Eq. 5.12 is valid, and there is

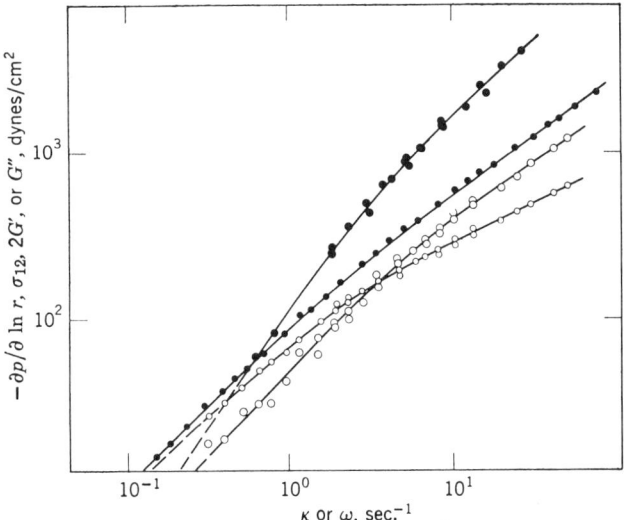

Fig. 5.14. Same as Fig. 5.13, but for polymethyl methacrylate in diethylphthalate.

substantial experimental support for Eq. 5.14, although data obtained over a wide range of $\dot{\gamma}$ and $\bar{\omega}$ indicate failure at high rates of deformation (38,39,40). The validity of Eq. 5.13 is less certain, simply because so few comparisons have been made. Figure 5.15 shows some values of $2G'/\bar{\omega}^2$, calculated at low frequency, where it approaches a constant value, plotted against $\eta_0 \lambda_1$, with λ_1 taken from the Bueche theory. If this graph is compared to Fig. 5.11 it is seen that the correlation is similar. However, the scatter in both of these figures is too great to allow an argument for Eq. 5.13, except in those cases for which a direct comparison has shown $\psi_{12}^{(0)}$ to equal the low frequency limit of $2G'/\bar{\omega}^2$.

Examination of Eq. 5.14 suffers primarily from a lack of availability of ψ_{12} data over an appreciable range of shear rate. The frequency dependence of G', however, was studied in a series of monodisperse polystyrenes by Harrison and coworkers (41). Molecular weights varied from 4.8×10^4 to 1.2×10^6. Dilute solutions in toluene were used. The data were correlated according to reduced variables, and a test of the Rouse-Zimm theory [Eq. 4.46] was presented. The data were in good agreement with the Zimm theory in the low frequency region, but shifted toward Rouse-like behavior

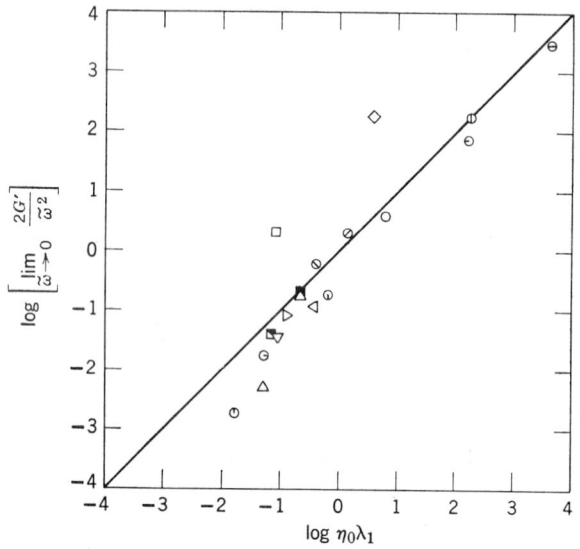

Fig. 5.15. Storage modulus at vanishing frequency for polymer solutions. Polystyrene: in toluene (38), (○) $M_v = 10^6$; in toluene (43), (◐) $M_v = 2 \times 10^6$; (⊖) $M_v = 2.5 \times 10^6$; in dioctyl phthalate (32), (⊘) $M_v = 5 \times 10^6$; in Aroclor (39), (⊗) $M_v = 1.2 \times 10^6$; in Aroclor (44), (△, ▽, ◁, ▷, ▲) $M_v = 2.7 \times 10^5$; in Aroclor (45), (◉) $M = 8 \times 10^4$, (⊖) $M = 2.4 \times 10^5$, (◎) $M = 4 \times 10^5$, (⊖) $M = 1.7 \times 10^6$. Polyisobutylene: in decalin (29), (□, ◊) $M_w = 10^6$; in Primol D (46), (■, ◩) $M = 8.4 \times 10^5$.

in the high frequency region. For the purpose of providing a comparison with normal stress behavior, the Rouse and Zimm theories have been replotted in Fig. 5.16 as

$$(2G'/\tilde{\omega}^2)/\lim_{\tilde{\omega} \to 0} (2G'/\tilde{\omega}^2) \text{ vs. } \lambda_1 \tilde{\omega}$$

with λ_1 given by Bueche's expression. The shading indicates the region in which Harrison's data fall. The curve representing the normal stress data of Fig. 5.12 is seen to differ from the storage modulus data in the region of large deformation rates. This is the same type of behavior as is evidenced, for example, in Fig. 5.14, and would seem to argue against the general validity of Eq. 5.15.

Any conclusions drawn from these results must be regarded as tentative, in view of the lack of extensive *comparative* data for well defined polymers. It appears that the $\eta(\dot{\gamma})$–$\eta'(\tilde{\omega})$ relationship is well established at low

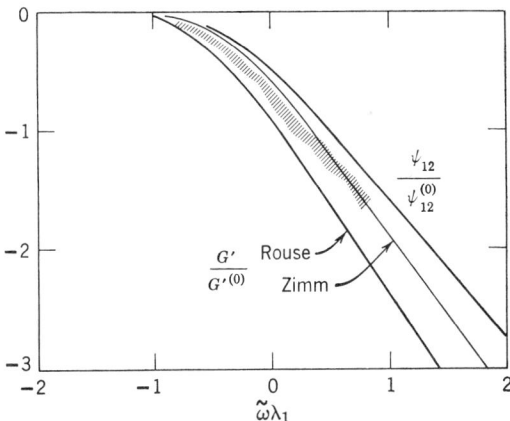

Fig. 5.16. Frequency dependence of storage modulus. Shaded area represents data for dilute solutions of monodisperse polystyrene in toluene (41). The Rouse and Zimm theories bracket the data. The curve through the normal stress data of Fig. 5.12 is shown for comparison.

deformation rates, but does not hold in all materials at high deformation rates. The low shear limiting values of ψ_{12} and $2G'/\tilde{\omega}^2$ appear to be identical in those materials for which direct comparison is available. At moderate shear rates ψ_{12} appears to be larger than $2G'/\tilde{\omega}^2$.

The degree to which these "analogies" hold is undoubtedly influenced by molecular weight distribution and, in solutions, by solvent effects ("hydrodynamic interaction" on a molecular scale). This latter point is reflected in the large degree of scatter observed when one attempts to correlate data obtained with different polymers, or different solvents, over a range of concentrations and molecular weights. A more detailed discussion of some aspects of the influence of the solvent on the storage modulus may be found in Ref. 44.

V. STRESS RELAXATION

In this chapter, thus far, consideration has been given primarily to properties normally associated with *fluids*. Theories of viscoelastic response, described in two earlier chapters, stress the importance of a "relaxation time" and suggest, in some cases, relationships between flow phenomena (such as the stresses accompanying shear flow) and stress relaxation. These relationships have been tested (and generally verified) in elastomeric materials [see, for example, Fig. 3.5], but only a few studies of stress relaxation in molten polymers or solutions have been reported.

Because fluids are incapable of holding their shape if unbounded, two methods have been developed which are well suited to the measurement of stress relaxation. In one technique, a steady shear flow is established so that the fluid is in a state of steady stress. The flow is suddenly stopped and the decay of this stress is followed. In a second type of experiment, fluid at rest is confined in the space between coaxial cylinders, or between a cone and plate. One surface is suddenly moved through a finite angle to a fixed position. Hence, the material is rapidly strained, and the relaxation of the accompanying stress, at constant strain, is followed.

An example of the former type of experiment is provided in the work of Schremp (48), who studied polymer solutions in a cone and plate system. Figure 5.17 shows his data, which indicate the effect of the initial shear rate on the rate at which the shear stress decays to equilibrium. The more rapid fall of the shear stress at the higher initial shear rate is a common observation, and is consistent with the predictions of the viscoelastic models presented in Eqs. 3.129 and 3.130, and Eqs. 3.76–3.79.

Bird and his coworkers (49) have presented more recent experimental studies of this type of stress relaxation phenomenon. The theoretical solution, for the models cited above, takes the form

$$\frac{\tau_{12}(t, \dot{\gamma}_0)}{\tau_{12}(0, \dot{\gamma}_0)} = \frac{\sum_{p=1}^{\infty} \frac{p^\alpha}{p^{2\alpha} + (c\lambda\dot{\gamma}_0)^2} e^{-p^\alpha t/\lambda}}{\sum_{p=1}^{\infty} \frac{p^\alpha}{p^{2\alpha} + (c\lambda\dot{\gamma}_0)^2}}$$

The parameters, c, α, and λ are defined in the equations cited above, and $\dot{\gamma}_0$ is the initial shear rate.

Two comments are appropriate here. Since c is of the order of unity, the term containing the initial shear rate is important only under such conditions that $\lambda\dot{\gamma}_0$ is of the order of unity or greater. It can be seen, also, that the relaxation behavior at long times is exponential decay, and that the decay rate is independent of $\dot{\gamma}_0$. Hence one could plot (as in Fig. 5.17) $\log[\tau_{12}(t, \dot{\gamma}_0)/\tau_{12}(0, \dot{\gamma}_0)]$ against t, and obtain the dominant relaxation time, λ, from the slope of the long time portion. Furthermore, this slope

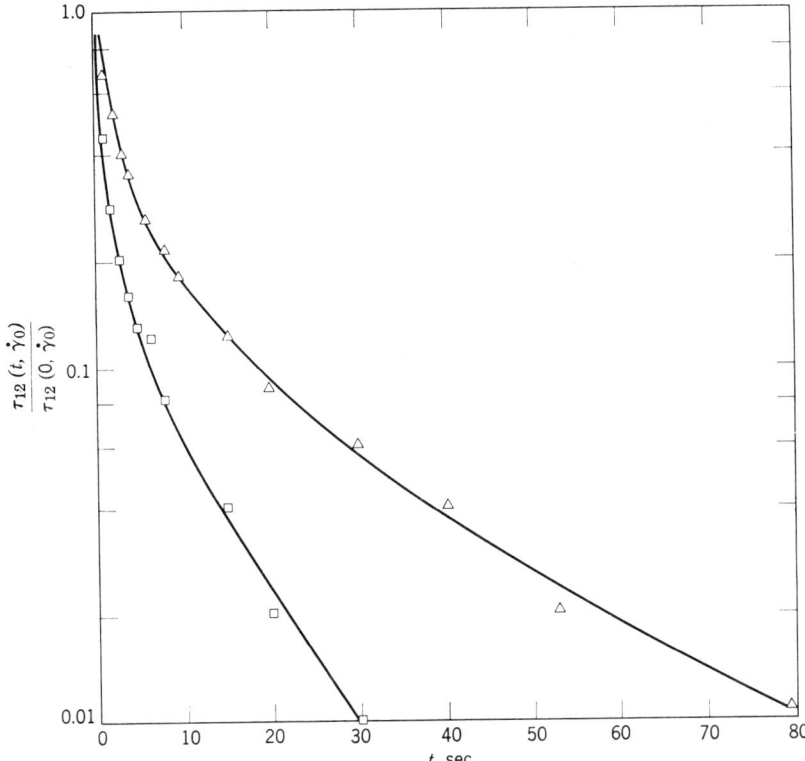

Fig. 5.17. Stress relaxation, upon cessation of steady shear, in a solution of polyisobutylene in decalin. $\dot{\gamma}_0$, sec:$^{-1}$. ($\triangle$) 0.087, ($\square$) 0.510. Data of Schremp, redrawn from Spriggs (48).

should be independent of $\dot{\gamma}_0$. Figure 5.17 appears to show data contrary to this expectation.

This point may be resolved by considering an estimate of the dominant relaxation time, using the Bueche theory. For this particular fluid, one calculates $\lambda_1 = 22$ sec. Even at the lowest shear rate ($\dot{\gamma}_0 = 0.087$ sec.$^{-1}$), the product $\lambda_1 \dot{\gamma}_0$ is greater than unity. Hence, both sets of data show a more rapid stress decay due to the influence of $\dot{\gamma}_0$. As a result, it seems likely that the stress falls to levels approaching the accuracy of the measuring system *before* the terminal decay period is reached. For example, the lowest measured stress for the lower curve is reached at $t = 30$ sec., so that t/λ_1 is not yet much larger than unity, and so is surely well short of the terminal region. Hence, one cannot take data, as presented in Fig. 5.17, and estimate the relaxation time of the fluid by using the Tobolsky-Murakami "Procedure X" [see p. 108]. If "Procedure X" had been used in this case, for example, the terminal relaxation time at $\dot{\gamma}_0 = 0.51$ sec.$^{-1}$ would have been calculated as 113 sec. (as compared to $\lambda_1 = 22$ sec.) and a similar estimate at $\dot{\gamma}_0 = 0.087$ sec.$^{-1}$ would yield $\lambda = 35$ sec.

Stress relaxation upon cessation of steady flow has also been studied in molten polymers (50,51). Because of the higher stress levels attainable, and the slower stress decay, melt relaxation is somewhat easier to measure than solution relaxation. An example of melt relaxation data is given by Ajroldi, Garbuglio, and Pezzin (50). Molten polystyrene was sheared in a cone-and-plate rheometer and the stress decay upon cessation of steady flow was measured. The stress decay curve was fitted with a sum of exponential decay terms. This is equivalent to the assumption that the material shows linear viscoelastic (maxwellian) response. The authors state that the steady shear rate preceding decay was in the newtonian flow region, and so linear relaxation may be a reasonable expectation.

The relaxation data were then used to calculate the viscosity vs. shear rate curve, which had been measured for the same material. The method used was based on an analysis of DeVries and Tochon (51), which adapts Eyring's rate theory (which leads to the Prandtl-Eyring viscosity model, Eq. 3.65) to the problem of stress relaxation. For a Maxwell model, stress relaxation upon cessation of steady flow would be given by Eq. 3.70. If a sum of maxwellian terms is considered, and if $G_p = \eta_p/\lambda_p$ is introduced, the result is

$$\tau_{12}(t, \dot{\gamma}_0) = \dot{\gamma}_0 \sum_{p=1}^{N} G_p \lambda_p \exp(-t/\lambda_p) \qquad (5.16)$$

DeVries and Tochon indicate that if one could measure a set of G_p and λ_p from stress relaxation data, the viscosity could be calculated from

$$\eta = (1/\dot{\gamma}) \sum_{p=1}^{N} G_p \text{ arc sinh}(\lambda_p \dot{\gamma}) \qquad (5.17)$$

Figure 5.18 shows a test of this idea (50) and indicates that the correspondence of theory and data is quite good. More extensive tests are shown in Reference 51 for other polymers.

If stress decay were dominated by a single relaxation time, then one would expect data to superpose on a plot of $\tau_{12}(t)/\tau_{12}(0)$ vs. t/λ, if the shear rate preceding decay were sufficiently low that it did not affect the decay. Vinogradov and Malkin (52) have tested superposition of stress decay data for molten polyisobutylene, polypropylene, and polyethylene, with the result shown in Fig. 5.19. They normalize the time variable with the zero shear viscosity (η_{max}, in their notation), and the correlation they achieve is simply a reflection of the degree to which η_0 is the dominant

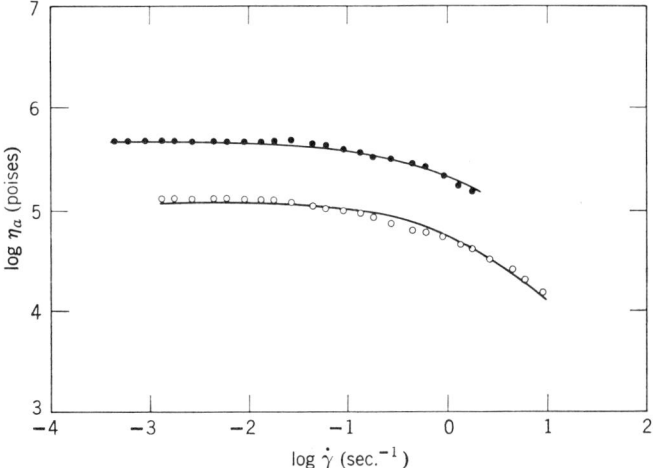

Fig. 5.18. Experimental apparent viscosity vs. shear rate for (○) polystyrene at 195°C. and (●) polymethylmethacrylate at 205°C. The solid lines are calculated from stress relaxation data by means of Eq. 5.17. Taken from Fig. 1 of Ref. 50.

variable affecting the relaxation time. The scatter in the correlation can be ascribed to the failure of η_0 to completely characterize the relaxation time, to the failure of a single relaxation time to accurately characterize the stress decay, and to the effect of molecular weight distribution on the material response.

The question of the dependence of the dominant relaxation time on material properties has already been discussed in examining the superposition of viscosity–shear rate data. Simple molecular theories, which predict $\lambda \sim \eta_0 M/\rho T$ (or $\eta_0 M/cT$, for solutions) are successful about as often as they fail. Other molecular theories (Williams, p. 160) suggest $\lambda \sim \eta_0/c^2 T$ in solutions, and supporting experiments have been cited. The effects of polymer–polymer interaction ("entanglement") and polymer–solvent interaction on dynamic moduli have been referred to, but corresponding studies of relaxation behavior do not exist.

The failure of a single relaxation time to characterize stress decay is evident from a plot such as Fig. 5.17, or Fig. 5.20, for a polyethylene melt (54), which also exhibits curvature in a semi-logarithmic plot. Figure 5.20 shows stress relaxation following the rapid imposition of a small shear strain in the sample. If the decay is taken to be maxwellian, so that

$$G(t) = \sum_{p=1}^{\infty} G_p e^{-t/\lambda_p},$$

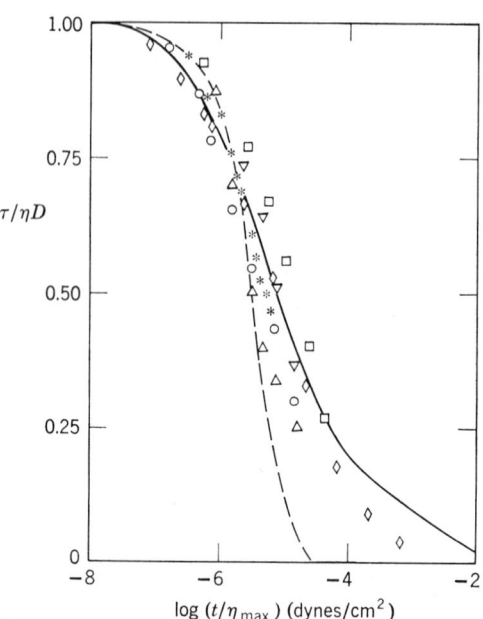

Fig. 5.19. Stress relaxation, upon cessation of steady shear, in molten polymers. (∗) Polyisobutylene, $M_w = 10^5$, $T = 22°C$. (○, □, △, ▽) Polypropylene, $M_w = 7 \times 10^5$, $T = 190–270°C$. (◊) Polyethylene data of Peticolas (53). $\tau/\eta D$ is identical to $\tau_{12}(t)/\tau_{12}(0)$.

then the Tobolsky-Murakami procedure may be used to obtain a discrete relaxation spectrum. Figure 5.20 shows the resultant spectrum, the height of the vertical lines representing the moduli. It can be seen that the relationship among the relaxation times does not follow the simple predictions of the Rouse theory (Eq. 4.60) or the Spriggs-Bird theory (Eq. 3.78). Furthermore, the moduli are not constant, as would be predicted, for example, by introducing $G_p = \eta_p/\lambda_p$ into Eq. 3.79.

The effect of molecular weight distribution on the relaxation of stress has been studied by Gogos and his coworkers (55). Figure 5.21 shows a comparison of the decay of stress, following the sudden imposition of a small strain, in two molten polystyrenes of the same weight average molecular weight ($M_w = 165{,}000$) but of different molecular weight distributions. The terminal relaxation time is larger in the broad distribution polymer.

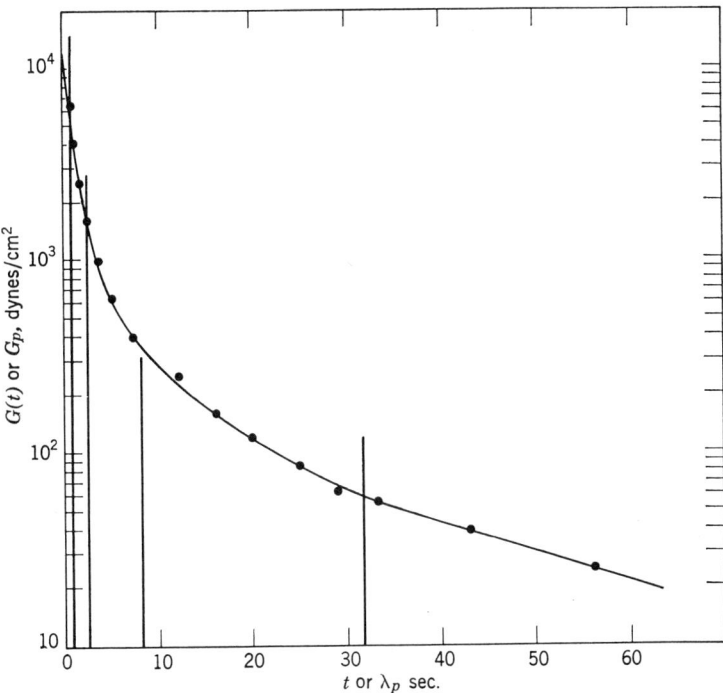

Fig. 5.20. Stress relaxation, at constant shear strain, in molten polyethylene (54). The data (●) give $G(t)$ vs. t. The vertical bars give G_p and λ_p.

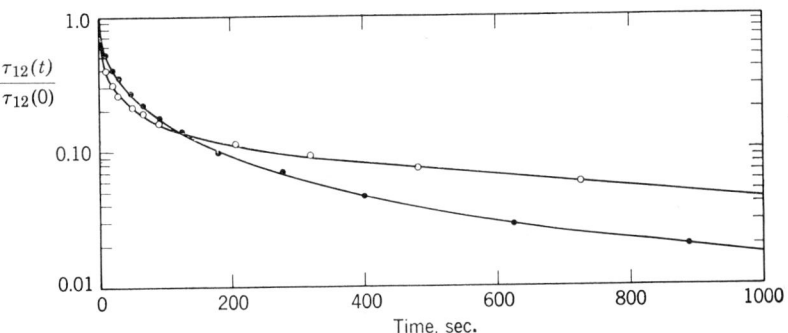

Fig. 5.21. Stress relaxation, at constant shear strain, in molten polystyrene. (●) Narrow molecular weight distribution. (○) Broad molecular weight distribution. Both materials have $M_w = 165{,}000$.

REFERENCES

1. Fox, T. G, and V. R. Allen, *J. Chem. Phys.*, **41**, 344 (1964); see also T. G Fox, *J. Polymer Sci. C*, **9**, 35 (1965).
2. Porter, R. S., and J. F. Johnson, *Proceedings of the Fourth International Congress on Rheology*, Pt. 2, E. H. Lee and A. L. Copley, Eds., Interscience, New York, 1965, p. 467.
3. Bueche, F., *Physical Properties of Polymers*, Interscience, New York, 1962.
4. Bueche, F., *J. Chem. Phys.*, **20**, 1959 (1952); **25**, 599 (1956).
5. Allen, V. R., and T. G Fox, *J. Chem. Phys.*, **41**, 337 (1964).
6. Ferry, J. D., *Viscoelastic Properties of Polymers*, Wiley, New York, 1961.
7. Simha, R., and R. F. Boyer, *J. Chem. Phys.*, **37**, 1003 (1962).
8. Johnson, M., W. Evans, I. Jordan, and J. D. Ferry, *J. Colloid Sci.*, **7**, 498 (1952).
9. Onogi, S., S. Kimura, K. Takashi, T. Masuda, and N. Miyanaga, *J. Polymer Sci. C*, **15**, 381 (1966).
10. Dunleavy, J., and S. Middleman, *Trans. Soc. Rheol.*, **10:1**, 157 (1966).
11. Ferry, J. D., L. Grandine, and D. Udy, *J. Colloid Sci.*, **8**, 529 (1953).
12. Ferry, J. D., E. Foster, G. Browning, and W. Sawyer, *J. Colloid Sci.*, **6**, 377 (1951).
13. Chinai, S., and W. Schneider, *J. Polymer Sci. A*, **3**, 1359 (1965).
14. Longworth, R., and W. Busse, *Trans. Soc. Rheol.*, **6**, 179 (1962).
15. Williams, M., *A.I.Ch.E. J.*, **13**, 534 (1967).
16. Teramoto, A., and H. Fujita, *Makromol. Chem.*, **85**, 261 (1965). See also Porter, R. S., and J. F. Johnson, *Trans. Soc. Rheol.*, **6**, 107 (1962).
17. Boyce, R. J., W. Bauer, and E. Collins, *Trans. Soc. Rheol.*, **10:2**, 545 (1967).
18. Bird, R. B., W. E. Stewart, and E. N. Lightfoot, *Transport Phenomena*, Wiley, New York, 1960, p. 28.
19. Williams, M. L., R. Landel, and J. D. Ferry, *J. Am. Chem. Soc.*, **77**, 3701 (1955).
20. Fox, T. G, and S. Loshaek, *J. Polymer Sci.*, **15**, 371 (1955); Fox, T. G, S. Gratch, and S. Loshaek, in *Rheology*, Vol. I, F. Eirich, Ed., Academic Press, New York, 1956, p. 450.
21. Wyman, D. P., L. J. Elyash, and W. J. Frazer, *J. Polymer Sci. A*, **3**, 681 (1965).
22. Ballman, R., and R. Simon, *J. Polymer Sci. A*, **2**, 3557 (1964).
23. Gruver, T., and G. Kraus, *J. Polymer Sci. A*, **2**, 797 (1964).
24. Holden, G., *J. Appl. Polymer Sci.*, **9**, 2911 (1965).
25. Vinogradov, G., and N. Prozorovskaya, *Rheol. Acta*, **3**, 156 (1964).
26. Ito, Y., *Bull. Chem. Soc. Japan*, **39**, 1368 (1966).
27. McKelvey, J. M., *Polymer Processing*, Wiley, New York, 1962.
28. Mendelson, R., *Trans. Soc. Rheol.*, **9:1**, 53 (1965).
29. DeWitt, T., H. Markovitz, F. Padden, and L. Zapas, *J. Colloid Sci.*, **10**, 174 (1955).
30. Seely, G., *A.I.Ch.E. J.*, **10**, 56 (1964).
31. Brodnyan, J., and E. Kelley, *Trans. Soc. Rheol.*, **5**, 205 (1961).
32. Tamura, M., M. Kurata, K. Osaki, and K. Tanaka, *J. Phys. Chem.*, **70**, 516 (1966).
33. LaNieve, H., and D. C. Bogue, paper presented to Society of Petroleum Engineers, Dec. 1966, Houston, Texas.
34. Ginn, R. F., and A. B. Metzner, *Proceedings of the Fourth International Congress on Rheology*, Pt. 2, E. H. Lee and A. L. Copley, Eds., Interscience, New York, 1965, p. 583.
35. Shertzer, C. R., and A. B. Metzner, *Proceedings of the Fourth International Congress on Rheology*, Pt. 2, E. H. Lee and A. L. Copley, Eds., Interscience, New York, 1965, p. 603.

36. Markovitz, H., and D. R. Brown, *Trans. Soc. Rheol.*, **7**, 137 (1963).
37. Kotaka, T., M. Kurata, and M. Tamura, *Rheol. Acta*, **2**, 179 (1962).
38. Osaki, K., M. Tamura, T. Kotaka, and M. Kurata, *J. Phys. Chem.*, **69**, 3642 (1965).
39. Tamura, M., M. Kurata, K. Osaki, and T. Katsuhisa, *J. Phys. Chem.*, **70**, 2271 (1966).
40. Onogi, S., T. Fujii, H. Kato, and S. Ogihara, *J. Phys. Chem.*, **68**, 1598 (1964).
41. Harrison, G., J. Lamb, and A. J. Matheson, *J. Phys. Chem.*, **68**, 1072 (1964).
42. Kotaka, T., M. Kurata, and M. Tamura, *J. Appl. Phys.*, **30**, 1705 (1959).
43. Osaki, K., M. Tamura, M. Kurata, and T. Kotaka, *J. Phys. Chem.*, **69**, 4183 (1965).
44. Holmes, L. A., K. Ninomiya, and J. D. Ferry, *J. Phys. Chem.*, **70**, 2714 (1966).
45. Tschoegl, N. W., and J. D. Ferry, *J. Phys. Chem.*, **68**, 867 (1964).
46. Frederick, J. E., N. W. Tschoegl, and J. D. Ferry, *J. Phys. Chem.*, **68**, 1974 (1964).
47. Leaderman, H., *J. Polymer Sci.*, **14**, 47 (1954).
48. Schremp, F. W., J. D. Ferry, and W. W. Evans, *J. Appl. Phys.*, **22**, 711 (1951); see also Spriggs, T. W., J. D. Huppler, and R. B. Bird, *Trans. Soc. Rheol.*, **10:1**, 191 (1966).
49. Huppler, J. D., I. F. Macdonald, E. Ashare, T. W. Spriggs, R. B. Bird, and L. A. Holmes, *Trans. Soc. Rheol.*, **11:2**, 181 (1967).
50. Ajroldi, G., C. Garbuglio, and G. Pezzin, *J. Appl. Polymer Sci.*, **11**, 289 (1967).
51. DeVries, A. J., and J. Tochon, *J. Appl. Polymer Sci.*, **7**, 315 (1963).
52. Vinogradov, G. V., and A. Ya. Malkin, *J. Polymer Sci. A-2*, **4**, 135 (1966).
53. Peticolas, W. L., *Rubber Chem. Technol.*, **36**, 1422 (1963).
54. Aloisio, C. J., S. Matsuoka, and B. Maxwell, *J. Polymer Sci. A-2*, **4**, 113 (1966).
55. Gogos, C., personal communication, June 1967.

Appendix A

Tensors

I. CARTESIAN TENSORS

We wish to describe the response of materials to external forces. In order to facilitate this description, it is necessary to develop a mathematics suitable to the quantities of interest. The method will be, first, to develop certain definitions and rules (logic) of this mathematics, then illustrate typical manipulations (operations) possible under the rules set up, and finally show by example that those quantities which we wish to characterize in rheology obey these rules and satisfy these definitions.

Before beginning an abstract mathematical discussion, however, it may be helpful to consider some physical concepts which appeal to intuition and experience, and show that these concepts are part of a more abstract mathematical framework.

A. Vectors

Let us consider, first a two-dimensional space (such as this sheet of paper) and two points P and O in that space, and ask "How does one denote the position of P with respect to O?" One might simply state that P is a distance R from O, but this is not sufficiently specific, for, as Fig. A.1 shows, many points (infinitely many) are a distance R from O.

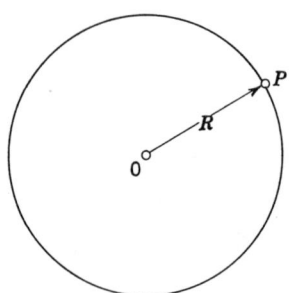

Figure A.1

TENSORS 201

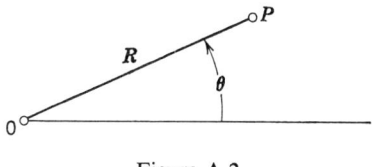

Figure A.2

One might draw (arbitrarily) some base line through the point O, as in Fig. A.2, and then state the distance R, and the angle the line OP makes counterclockwise from this reference line, say θ. This completely specifies the position of point P with respect to O. Of course it is recognized that this (R, θ) description is in what is commonly called a "polar coordinate system."

Alternately, as in Fig. A.3, one might draw a cartesian coordinate system with its origin at O, and give the coordinates (x_1, x_2), and again completely specify P with respect to O.

Any description of position in a two-dimensional space requires only two independent numbers. Thus if (x_1, x_2) and (R, θ) describe the same point P with respect to O, then of these *four* numbers, only two are independent of each other. We know, for example, that

$$R^2 = x_1^2 + x_2^2 \tag{A.1}$$

$$\theta = \tan^{-1}(x_2/x_1) \tag{A.2}$$

It should be obvious that the description of P depends upon the way the reference coordinate system is drawn. For example, we could have used a different coordinate system* and specified P by giving coordinates x_1' and

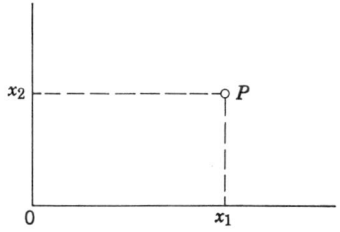

Figure A.3

* Unless otherwise stated, we now use only a cartesian coordinate system.

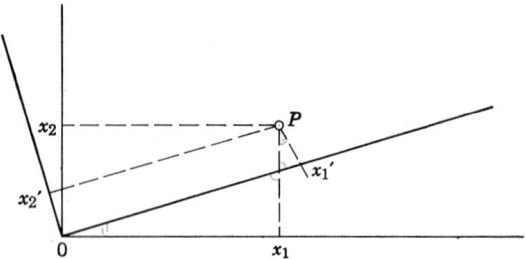

Figure A.4

x_2', as in Fig. A.4. Since this specification seems arbitrary, in the sense that it depends upon an arbitrary choice of coordinate system, we might inquire as to whether any general relationships exist which allow us to transform the position specification from one coordinate system to another, if we know the relationship between the coordinate systems.* We might also inquire as to whether there is anything about the specification of position which does *not* depend on our choice of coordinate system.

To answer the first question we note that, from Fig. A.4

$$x_1 = x_1' \cos(x_1, x_1') - x_2' \sin(x_1, x_1') \tag{A.3}$$

$$x_2 = x_1' \sin(x_1, x_1') + x_2' \cos(x_1, x_1') \tag{A.4}$$

where $\sin(x_1, x_1')$ means the sine of the angle traversed in moving from the line ox_1 to the line ox_1'.

We will rearrange this pair of equations somewhat for a purpose which will become clear soon. Note first that

$$\sin(x_1, x_1') = -\cos(x_1, x_2') = \cos(x_2, x_1') \tag{A.5}$$

and

$$\cos(x_1, x_1') = \cos(x_2, x_2') \tag{A.6}$$

Let us define

$$a_{11} = \cos(x_1, x_1') \tag{A.7}$$

$$a_{21} = \cos(x_2, x_1') \tag{A.8}$$

$$a_{12} = \cos(x_1, x_2') \tag{A.9}$$

$$a_{22} = \cos(x_2, x_2') \tag{A.10}$$

* Equations A.1 and A.2 are such a relationship between polar and cartesian coordinates.

Then we find that Eqs. A.3 and A.4 become

$$x_1 = x'_1 a_{11} + x'_2 a_{12} \tag{A.11}$$

$$x_2 = x'_1 a_{21} + x'_2 a_{22} \tag{A.12}$$

We can easily verify that the above equations can be represented by

$$x_i = \sum_{k=1}^{2} a_{ik} x'_k \qquad i = 1, 2 \tag{A.13}$$

where

$$a_{ik} = \cos(x_i, x'_k) \tag{A.14}$$

If we adopt the convention that whenever an index is repeated in one term a summation is implied, then we may write

$$x_i = a_{ik} x'_k \qquad i = 1, 2 \tag{A.15}$$

By similar arguments we may show that

$$x'_k = a_{ik} x_i \qquad k = 1, 2 \tag{A.16}$$

This pair of equations constitutes a *transformation rule*. It tells us how to relate the x_i of one coordinate system to the x'_i of another, given the relationship between the systems (a_{ik}).

It is possible to show that in three dimensions Eqs. A.15 and A.16 hold, provided we sum k or i from 1 to 3.

We note that the index over which summation is taken (i in Eq. A.16) is a "dummy" index, and could be represented by any index, with the exception of indices which already appear in the equation. Equation A.16 could be written as

$$x'_k = a_{lk} x_l$$

but not as

$$x'_k = a_{kk} x_k$$

We now state that if any numbers x_i, defined in one coordinate system, are related to other numbers x'_k in another coordinate system by the pair of transformations written above (A.15 and A.16), these numbers are the components of a *vector*. Thus, we have *defined a vector mathematically* based on its rule of transformation, and we have shown that the position of one point of a medium relative to another is given by a vector, defined in some coordinate system.

Now let us illustrate some formal operations with vector components.

Consider $x_i = a_{ik} x'_k$, and its "inverse," $x'_i = a_{li} x_l$, or, since i is a dummy index, $x'_k = a_{lk} x_l$. We can differentiate the first expression with respect to x_j:

$$\frac{\partial x_i}{\partial x_j} = a_{ik} \frac{\partial x'_k}{\partial x_j} = a_{ik} a_{lk} \frac{\partial x_l}{\partial x_j} \tag{A.17}$$

Since x_i and x_j are independent variables

$$\frac{\partial x_l}{\partial x_j} = 0 \tag{A.18}$$

unless $l = j$, in which case

$$\frac{\partial x_l}{\partial x_j} = \frac{\partial x_l}{\partial x_l} = 1 \tag{A.19}$$

We can write Eqs. A.18 and A.19 as a single equation, by defining

$$\delta_{ij} = \begin{cases} 0 & \text{if} \quad i \neq j \\ 1 & \text{if} \quad i = j \end{cases} \tag{A.20}$$

Then

$$\frac{\partial x_l}{\partial x_j} = \delta_{lj} \tag{A.21}$$

δ_{lj} is called the Kronecker delta.

Now we have

$$\delta_{ij} = a_{ik}a_{lk}\delta_{lj} \tag{A.22}$$

We can express all of the equations represented by A.22 succinctly by noticing that δ_{lj} is zero unless $l = j$. Hence

$$\delta_{ij} = a_{ik}a_{jk} \tag{A.23}$$

Now let us attempt to put this formalism to work for us. Consider, for example

$$x_i = a_{ik}x'_k$$

and

$$x_j = a_{jk}x'_k$$

Then

$$x_i x_j = a_{ik}a_{jk}x'_k x'_k = \delta_{ij}x'_k x'_k$$

or

$$x_i x_i = x'_k x'_k \tag{A.24}$$

By the summation convention

$$x_1 x_1 + x_2 x_2 + x_3 x_3 = x'_1 x'_1 + x'_2 x'_2 + x'_3 x'_3$$

or

$$x_1^2 + x_2^2 + x_3^2 = {x'_1}^2 + {x'_2}^2 + {x'_3}^2 \tag{A.25}$$

This equation states that the sum of the squares of the components of a vector in one coordinate system is equal to the corresponding sum in another coordinate system; in other words, the sum of the squares of components of a vector is *independent of coordinate system*. This answers a previously posed question, as to whether *any* properties of the specification of position in space were independent of the choice of coordinate system.

Now we will recognize that this particular property is just the square of the *distance* between points, which certainly is independent of how we arbitrarily set up a coordinate system.

The formalism which led us to this point is surely more tedious than the intuitive geometrical arguments which would have led us to the same point. We will find, however, that such concrete intuitive arguments are lacking when we consider more complex properties of materials, and that the mathematical formalism provides an avenue to analysis which is shorter than any other approach.

B. Cartesian Tensors

We have defined a vector based upon its transformation property. We now state this definition precisely:

Any quantity A having three components A_i referred to a set of cartesian axes $x_1 x_2 x_3$ is a vector, or a *tensor of the first rank*, if its components A'_i referred to a rotated set of cartesian axes $x'_1 x'_2 x'_3$ satisfy the relations

$$A'_i = a_{ji} A_j \tag{A.26}$$

or

$$A_i = a_{ij} A'_j \tag{A.27}$$

We now extend this definition formally to define a *second-rank tensor*:

Any quantity $\mathbf{T}$ having nine components T_{ij} referred to a set of cartesian axes $x_1 x_2 x_3$ is a *tensor of the second rank* if its components T'_{ij} referred to a rotated set of cartesian axes $x'_1 x'_2 x'_3$ satisfy the relations

$$T'_{ij} = a_{ki} a_{lj} T_{kl} \tag{A.28}$$

or

$$T_{ij} = a_{ik} a_{jl} T'_{kl} \tag{A.29}$$

Suppose we have a tensor $\mathbf{T}$ and a vector A, and we consider the operation $\mathbf{T}A$, defined as the sum $T_{ij} A_j$. We sum over j and get, in general, a different number for each choice of i. We call this sum $B(i)$:

$$B(i) = T_{ij} A_j \tag{A.30}$$

Now, if the components of $\mathbf{T}$ and A were given in another coordinate system, we would write

$$B'(i) = T'_{ij} A'_j \tag{A.31}$$

Since $\mathbf{T}$ is a tensor, and A a vector,

$$B'(i) = (T_{kl} a_{ki} a_{lj})(a_{mj} A_m) = a_{ki}(a_{lj} a_{mj}) T_{kl} A_m \tag{A.32}$$

We have seen previously that

$$\delta_{lm} = a_{lj} a_{mj}$$

Hence
$$B'(i) = a_{ki}T_{kl}A_m\delta_{lm} = a_{ki}T_{kl}A_l \qquad (A.33)$$
or
$$B'(i) = a_{ki}B(k) \qquad (A.34)$$
But, except for notation, this is just the transformation rule for a vector:
$$B'_i = a_{ki}B_k$$
Thus we have shown that $B(i)$ is a vector, or that $T_{ij}A_j$ produces a vector
$$B_i = T_{ij}A_j \qquad (A.35)$$
or, in operator notation $B = TA$.

We interpret this equation in the following manner: given a vector A and a tensor T, the operation TA, defined above, produces another vector B, in general, of different magnitude and direction from A.

We can illustrate tensor multiplication with a two-dimensional case. We will arrange the components of a tensor T in an array:
$$T = \begin{pmatrix} T_{11} & T_{12} & T_{13} \\ T_{21} & T_{22} & T_{23} \\ T_{31} & T_{32} & T_{33} \end{pmatrix} \qquad (A.36)$$

We denote the components of a vector by $A = (A_1, A_2, A_3)$.

Let $T = \begin{pmatrix} 1 & 2 & 0 \\ -1 & 3 & 0 \\ 0 & 0 & 0 \end{pmatrix}$ and $A = (1, 1, 0)$

Then $B = (T_{11}A_1 + T_{12}A_2 + T_{13}A_3, T_{21}A_1 + T_{22}A_2 + T_{23}A_3,$
$T_{31}A_1 + T_{32}A_2 + T_{33}A_3)$
or $\quad B = (1 + 2, -1 + 3, 0) = (3, -2, 0)$.

Obviously B and A are of different magnitude and direction.

A tensor is a linear operator, i.e.
$$T(A + B) = TA + TB \qquad (A.37)$$
and
$$T(cA) = cTA \ (c \text{ a constant}) \qquad (A.38)$$

It is easily shown that the sum of two tensors is a tensor.

A tensor T is said to be symmetric if $T_{ij} = T_{ji}$ and antisymmetric if $T_{ij} = -T_{ji}$.

By noting that any tensor T may be written
$$T_{ij} = \tfrac{1}{2}(T_{ij} + T_{ji}) + \tfrac{1}{2}(T_{ij} - T_{ji})$$

and observing that $T_{ij} + T_{ji}$ is a symmetric tensor, and $T_{ij} - T_{ji}$ is an antisymmetric tensor, it is clear that *any* tensor may be written as the sum of a symmetric and antisymmetric tensor.

For example

$$\begin{pmatrix} 1 & 4 & 7 \\ 2 & 5 & 8 \\ 3 & 6 & 9 \end{pmatrix} = \frac{1}{2}\begin{pmatrix} 2 & 6 & 10 \\ 6 & 10 & 14 \\ 10 & 14 & 18 \end{pmatrix} + \frac{1}{2}\begin{pmatrix} 0 & 2 & 4 \\ -2 & 0 & 2 \\ -4 & -2 & 0 \end{pmatrix}$$

$$= \begin{pmatrix} 1 & 3 & 5 \\ 3 & 5 & 7 \\ 5 & 7 & 9 \end{pmatrix} + \begin{pmatrix} 0 & 1 & 2 \\ -1 & 0 & 1 \\ -2 & -1 & 0 \end{pmatrix}$$

It should be clear that in general the diagonal elements of an antisymmetric tensor are zero.

A particularly simple symmetric tensor is a *diagonal* tensor:

$$\mathbf{D} = \begin{pmatrix} D_{11} & 0 & 0 \\ 0 & D_{22} & 0 \\ 0 & 0 & D_{33} \end{pmatrix} \tag{A.39}$$

a tensor all of whose elements are zero except those along the main diagonal.

Note that, consistent with its definition (Eq. A.20)

$$\delta_{ij} = \begin{pmatrix} 1 & 0 & 0 \\ 0 & 1 & 0 \\ 0 & 0 & 1 \end{pmatrix} \tag{A.40}$$

The fact that the Kronecker delta may be written in the form of an array does not make it a tensor. It is not difficult to show, however, that $\boldsymbol{\delta}$ *is* a second-rank tensor. It is called an *isotropic* tensor, since, by definition, it has the same components in all coordinate systems.

We now ask whether there are any circumstances under which, given the *symmetric* tensor **T**, there is a vector A such that the tensor multiplication $\mathbf{T}A$ produces a vector with the same direction as A, say λA, where λ is some constant. Equivalently, we ask for the solution of the equation $\mathbf{T}A = \lambda A$. We state here that we shall attempt to answer this question by giving the components of the desired vector A_i with respect to the coordinate system in which have been defined the T_{ij}. It should be clear that there is no loss of generality if we take A to be a unit vector.

APPENDIX A

The tensor equation $\mathbf{T}A = \lambda A$ is equivalent, for $\mathbf{T}$ a symmetric tensor, to the three scalar equations

$$T_{11}A_1 + T_{12}A_2 + T_{13}A_3 = \lambda A_1$$
$$T_{12}A_1 + T_{22}A_2 + T_{23}A_3 = \lambda A_2$$
$$T_{13}A_1 + T_{23}A_2 + T_{33}A_3 = \lambda A_3$$

or

$$(T_{11} - \lambda)A_1 + T_{12}A_2 + T_{13}A_3 = 0 \qquad (A.41)$$
$$T_{12}A_1 + (T_{22} - \lambda)A_2 + T_{23}A_3 = 0 \qquad (A.42)$$
$$T_{13}A_1 + T_{23}A_2 + (T_{33} - \lambda)A_3 = 0 \qquad (A.43)$$

We must solve this set of three simultaneous equations for the components A_1, A_2, A_3. Obviously $A_1 = A_2 = A_3 = 0$ is a (trivial) solution. The condition that there are solutions other than the trivial solution is that the determinant of the coefficients must vanish.

Thus we have the *secular equation*

$$\det(\lambda) = \begin{vmatrix} T_{11} - \lambda & T_{12} & T_{13} \\ T_{12} & T_{22} - \lambda & T_{23} \\ T_{13} & T_{23} & T_{33} - \lambda \end{vmatrix} = 0 \qquad (A.44)$$

We use the functional notation $\det(\lambda)$ as a reminder that, since the T_{ij} are assumed known, the determinant is a function of λ.

We now have a cubic equation to solve for λ, of the form

$$\lambda^3 - I_T\lambda^2 + II_T\lambda - III_T = 0 \qquad (A.45)$$

where

$$I_T = T_{11} + T_{22} + T_{33} \qquad (A.46)$$

$$II_T = T_{11}T_{22} + T_{22}T_{33} + T_{33}T_{11} - T_{12}^2 - T_{13}^2 - T_{23}^2 \qquad (A.47)$$

and

$$III_T = T_{11}T_{22}T_{33} + 2T_{12}T_{23}T_{13} - T_{11}T_{23}^2 - T_{22}T_{13}^2 - T_{33}T_{12}^2 \qquad (A.48)$$

When T_{ij} is symmetric, Eq. A.45 has three real roots, say $\lambda^{(1)}$, $\lambda^{(2)}$, and $\lambda^{(3)}$. If we choose one such root, $\lambda^{(1)}$, and insert it into the Eqs. A.41 to A.43, then we can solve from any two of these equations for the ratios A_1/A_3, A_2/A_3. Finally, with the equation $A_1^2 + A_2^2 + A_3^2 = 1$ (since A is a unit vector) we can solve for A_1, A_2, and A_3.

Example: Let us take a two-dimensional symmetric tensor

$$T = \begin{pmatrix} 1 & 2 & 0 \\ 2 & 4 & 0 \\ 0 & 0 & 0 \end{pmatrix}$$

Equation A.45 becomes $\lambda^3 - 5\lambda^2 = 0$, so that $\lambda^{(1)} = \lambda^{(2)} = 0$ and $\lambda^{(3)} = 5$.

If $\lambda^{(3)}$ is now substituted into Eq. A.43, we find $A_3 = 0$, while, if $\lambda^{(3)}$ is substituted into either Eq. A.41 or A.42, we find $A_1/A_2 = \frac{1}{2}$.

Since A is assumed a unit vector, we know that $A_1^2 + A_2^2 = 1$ and it follows immediately that

$$A_1 = 1/\sqrt{5}$$
$$A_2 = 2/\sqrt{5}$$

or

$$A^{(3)} = (1/\sqrt{5}, 2/\sqrt{5}, 0)$$

We have used a superscript (^{3}) on A to remind us that this is the solution obtained using the value $\lambda^{(3)}$. If $\lambda^{(1)}$ is used, then a different solution is found, which is

$$A^{(1)} = A^{(2)} = (-2/\sqrt{5}, 1/\sqrt{5}, 0)$$

It can be easily verified that $\mathbf{T}A = \lambda A$ for either of these vectors A.

We note that, in general, for each $\lambda^{(i)}$ we insert into Eqs. A.41 to A.43, we obtain a different set of $A^{(i)}$. Thus in general, there are three distinct vectors A such that $\mathbf{T}A = \lambda A$.

We shall now show that these three vectors are mutually orthogonal, and therefore may be used to define a cartesian coordinate system. Then we demonstrate the point of this exposition; that a symmetric tensor $\mathbf{T}$, when referred to a coordinate system defined by the vectors $A^{(i)}$ such that $\mathbf{T}A = \lambda A$, is a diagonal tensor with components

$$\mathbf{T'} = \begin{pmatrix} \lambda^{(1)} & 0 & 0 \\ 0 & \lambda^{(2)} & 0 \\ 0 & 0 & \lambda^{(3)} \end{pmatrix} \quad (A.49)$$

Let the vector associated with the root $\lambda^{(i)}$ be $A^{(i)}$, with components $(A_1^{(i)}, A_2^{(i)}, A_3^{(i)})$, as in Fig. A.5. Now, since we have taken A to be a unit vector, the numbers $A_j^{(i)}$ are really the direction cosines between the 1, 2, 3 system and the $A^{(i)}$ vector. In other words, the $A_j^{(i)}$ are really the numbers a_{ji} we have used before.

Now the vector A has been determined so that $\mathbf{T}A = \lambda A$, or, in component form

$$T_{ij}A_j^{(i)} = \lambda^{(i)}A_i^{(i)} \quad \text{[No sum implied over } (i).\text{]} \quad (A.50)$$

Writing this equation for each root $\lambda^{(i)}$ we have

$$T_{ij}A_j^{(1)} = \lambda^{(1)}A_i^{(1)} \quad (A.51)$$
$$T_{ij}A_j^{(2)} = \lambda^{(2)}A_i^{(2)} \quad (A.52)$$
$$T_{ij}A_j^{(3)} = \lambda^{(3)}A_i^{(3)} \quad (A.53)$$

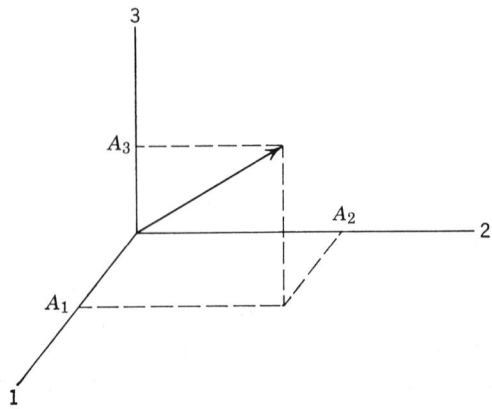

Figure A.5

Multiply the first equation by $A_i^{(2)}$ and the second by $A_i^{(1)}$.

$$T_{ij}A_j^{(1)}A_i^{(2)} = \lambda^{(1)}A_i^{(1)}A_i^{(2)} \tag{A.54}$$

$$T_{ij}A_j^{(2)}A_i^{(1)} = \lambda^{(2)}A_i^{(2)}A_i^{(1)} \tag{A.55}$$

Since we sum over i and j on the left-hand side of both equations, i and j are dummy indices and can be changed. In particular, they can be reversed, to give, in the second equation

$$T_{ji}A_j^{(1)}A_i^{(2)} = \lambda^{(2)}A_i^{(2)}A_i^{(1)} \tag{A.56}$$

(There is no need to change i to j on the right-hand side.)

Since we have assumed **T** to be a symmetric tensor, $T_{ij} = T_{ji}$, the left-hand sides of Eqs. A.54 and A.56 are seen to be equal, and we find

$$(\lambda^{(1)} - \lambda^{(2)})A_i^{(1)}A_i^{(2)} = 0 \tag{A.57}$$

If $\lambda^{(1)} \neq \lambda^{(2)}$, then

$$A_i^{(1)}A_i^{(2)} = 0 \tag{A.58}$$

But $A_i^{(1)}A_i^{(2)}$ is just the scalar product of $A^{(1)}$ and $A^{(2)}$, so that we have shown $A^{(1)} \cdot A^{(2)} = 0$. Hence we have shown that the vectors $A^{(1)}$ and $A^{(2)}$, are perpendicular, and the same may be shown for $A^{(1)}$ and $A^{(3)}$, and $A^{(2)}$ and $A^{(3)}$.

This demonstrates that the three vectors $A^{(i)}$ are mutually orthogonal, and so may be used to define a cartesian coordinate system. Such a set of axes is called "the principal axes of the symmetric tensor **T**."

Now, with respect to the principal axes, by the definition of a tensor, we have

$$T'_{kl} = a_{ik}a_{jl}T_{ij}$$

But we have seen, in Eq. A.50, with (l) for (i), that $T_{ij}a_{jl} = \lambda^{(l)}a_{il}$. (We have used the fact that $A_j^{(l)} = a_{jl}$.) Then

$$T'_{kl} = a_{ik}a_{il}\lambda^{(l)} = \lambda^{(l)}\,\delta_{kl} \tag{A.59}$$

and

$$T'_{kl} = \begin{cases} 0 & \text{if } k \neq l \\ \lambda^{(l)} & \text{if } k = l \end{cases}$$

or

$$T'_{kl} = \begin{pmatrix} \lambda^{(1)} & 0 & 0 \\ 0 & \lambda^{(2)} & 0 \\ 0 & 0 & \lambda^{(3)} \end{pmatrix} \tag{A.60}$$

In summary, if we are given a *symmetric* tensor **T**, we can always find a coordinate system (defined by three orthogonal vectors $A^{(i)}$) such that, in this new coordinate system, **T** becomes a diagonal tensor. The virtue of this fact is that we can often prove tensor relationships in the principal coordinate system with relative simplicity, and then transform formally to a general coordinate system to express the general tensor equation.*

Example

$$T = \begin{pmatrix} 3 & 0 & 1 \\ 0 & 3 & -2 \\ 1 & -2 & 7 \end{pmatrix} \quad \text{Find the principal axes of } \mathbf{T}.$$

$$\det(\lambda) = \begin{vmatrix} 3-\lambda & 0 & 1 \\ 0 & 3-\lambda & -2 \\ 1 & -2 & 7-\lambda \end{vmatrix}$$

$$= (3-\lambda)[(3-\lambda)(7-\lambda) - 4] + \lambda - 3 = 0$$

$$\lambda^{(1)} = 3 \quad \lambda^{(2)} = 8 \quad \lambda^{(3)} = 2$$

$$(3-\lambda)A_1 + A_3 = 0$$
$$(3-\lambda)A_2 - 2A_3 = 0$$
$$A_1 - 2A_2 + (7-\lambda)A_3 = 0$$

$$\lambda^{(1)} = 3 \quad \lambda^{(2)} = 8 \quad \lambda^{(3)} = 2$$

---

* An example is the Cayley-Hamilton theorem: $\mathbf{T}^3 - I_T\mathbf{T}^2 + II_T\mathbf{T} - III_T = 0$ (A.45a). This follows for **T'**, defined by Eq. A.49, since the components of **T'** obey Eq. A.45. An equation valid in the principal coordinate system is valid in any coordinate system.

$$A_3 = 0 \qquad\qquad -5A_1 + A_3 = 0 \qquad\qquad A_1 + A_3 = 0$$
$$A_1 = 2A_2 \qquad\qquad -5A_2 - 2A_3 = 0 \qquad\qquad A_2 - 2A_3 = 0$$
$$\qquad\qquad\qquad A_1 - 2A_2 - A_3 = 0 \qquad\qquad A_1 - 2A_2 + 5A_3 = 0$$
$$\qquad\qquad\qquad A_1^2 + A_2^2 + A_3^2 = 1$$

$$A_1^{(1)} = 2/\sqrt{5} = a_{11} \qquad A_1^{(2)} = 1/\sqrt{30} = a_{12} \qquad A_1^{(3)} = -1/\sqrt{6} = a_{13}$$
$$A_2^{(1)} = 1/\sqrt{5} = a_{21} \qquad A_2^{(2)} = -2/\sqrt{30} = a_{22} \qquad A_2^{(3)} = 2/\sqrt{6} = a_{23}$$
$$A_3^{(1)} = 0 = a_{31} \qquad A_3^{(2)} = 5/\sqrt{30} = a_{32} \qquad A_3^{(3)} = 1/\sqrt{6} = a_{33}$$

It is not difficult to transform **T** to the principal coordinate system and verify that

$$\mathbf{T}' = \begin{pmatrix} 3 & 0 & 0 \\ 0 & 8 & 0 \\ 0 & 0 & 2 \end{pmatrix}$$

We now inquire whether any functions of the components of a tensor are independent of the choice of reference frame in which the components are given. If such a function can be found, for a given tensor, it is called an *invariant* of that tensor. We have already found an invariant of a first-order tensor, or vector, namely, $x_i x_i$ of Eq. A.24.

Let us consider first the quantity T_{ii} of *any* second-order tensor. By definition

$$T'_{ii} = a_{ki} a_{ji} T_{kj} \tag{A.61}$$

From Eq. A.23, we find immediately

$$T'_{ii} = \delta_{kj} T_{kj} = T_{kk} = T_{ii} \tag{A.62}$$

Hence $T_{ii} = T_{11} + T_{22} + T_{33}$ is an invariant of the tensor **T**. Note that T_{ii} is just I_T of Eq. A.46.

It can be verified that II_T of Eq. A.47 is a special case, for a symmetric tensor, of the more general expression

$$II_T = \tfrac{1}{2}(T_{ii} T_{jj} - T_{ij} T_{ij}) \tag{A.63}$$

We have just shown T_{ii} is invariant, and it follows immediately that $T_{ii} T_{jj}$ is also invariant.

Now we know that

$$T_{ij} T_{ij} = (a_{ik} a_{jm} T'_{km})(a_{ik} a_{jm} T'_{km})$$

But we must be careful here, for we see that the dummy indices k and m appear *four* times in the right-hand side of this equation. This must be avoided, and is easily done by changing one set of indices so that

$$T_{ij} T_{ij} = (a_{ik} a_{jm} T'_{km})(a_{ir} a_{js} T'_{rs}) \tag{A.64}$$

Now it is simple to show that

$$T_{ij}T_{ij} = (a_{ik}a_{ir})(a_{jm}a_{js})T'_{km}T'_{rs} = \delta_{kr}\delta_{ms}T'_{km}T'_{rs}$$
$$= T'_{km}T'_{km} = T'_{ij}T'_{ij} \qquad (A.65)$$

The invariance of II_T now follows from the fact that the sum of two invariants is itself invariant.

Finally we remark that III_T of Eq. A.48 is a special case, for a symmetric tensor, of the more general definition

$$III_T = \epsilon_{ijk}T_{1i}T_{2j}T_{3k} \qquad (A.66)$$

where ϵ_{ijk} is the "permutation symbol," defined so that

$$\epsilon_{ijk} = \begin{cases} 0 & \text{if any two indices are equal} \\ +1 & \text{if the indices are cyclic (123, 231, or 312)} \\ -1 & \text{if the indices are acyclic (132, 213, or 321)} \end{cases}$$

We state without giving a proof that III_T is an invariant of a second-order tensor.

I_T, II_T, and III_T are called the first, second, and third invariants, respectively, of the tensor **T**. We note that, although the invariants were originally introduced in a discussion of a *symmetric* tensor, they are here defined for *any* second-order tensor.

II. CURVILINEAR COORDINATES

Thus far our discussion has been restricted to cartesian tensors. This is a restriction only in convenience, for, since a cartesian coordinate system can always be established, mechanics and dynamics, and so rheology, can always be completely described in cartesian terms.

In certain types of problems one finds that curvilinear coordinates are far more convenient to use than cartesian coordinates. Hence we are motivated to develop some information regarding the mathematical behavior of tensors defined in curvilinear coordinate systems. Of necessity, much of the development will be quite general, and will appear at times to be far removed from physical application. Before embarking on this mathematical excursion, then, we offer a simple example to indicate that our knowledge of cartesian *vectors* is inadequate to discuss even simple physics in polar coordinates.

Consider a mass point traveling in a plane circular orbit with constant angular velocity. We wish to describe the position **P**, velocity **v**, and acceleration **a**, of the particle. We assume **P**, **v**, and **a** are given in a cartesian coordinate system. We desire **P**′, **v**′, and **a**′ in a plane polar coordinate system related to the cartesian system by

$$x = r \cos \theta \quad \text{or} \quad \begin{array}{l} r = \sqrt{x^2 + y^2} \\ \theta = \tan^{-1}(y/x) \end{array}$$
$$y = r \sin \theta$$

Let the constant angular velocity be $\omega = d\theta/dt$ and let the constant radius of the orbit be $r = R$.

If **P** is given, then, by the transformation of coordinates given above, **P**' follows immediately:

$$\mathbf{P}' = (r, \theta) = (\sqrt{x^2 + y^2}, \tan^{-1}(y/x))$$

The velocity vector in cartesian coordinates has components $\mathbf{v} = (dx/dt, dy/dt)$.

In polar coordinates the velocity is $\mathbf{v}' = (dr/dt, d\theta/dt) = (0, \omega)$. The velocity vector has only an angular component, consistent with our assumption that the orbit is one of constant radius.

The acceleration vector in a cartesian system is $a = (d^2x/dt^2, d^2y/dt^2)$.

At first glance it might seem that for polar coordinates we should define the components of acceleration to be the time derivatives of the components of velocity, since this is true for cartesian coordinates. The result would be

$$\mathbf{a}' = (d^2r/dt^2, d^2\theta/dt^2) = (0, 0)$$

which is clearly incorrect. Anyone who recalls some physics will realize that we have left out the centripetal acceleration, and that **a**' is given correctly by $\mathbf{a}' = (-R\omega^2, 0)$.

But suppose we did not know anything about the physics of acceleration. After all, we are going to be investigating the behavior of materials; we cannot use what we are yet to discover to direct our investigation. In that case the reader might be willing to recall that a vector possesses not only a magnitude, but a *direction* as well. Thus, while the magnitude of the velocity vector (speed) is unchanged in this example, its direction is continually changing. As a consequence of this observation, which really follows from a definition of the vector, and so is independent of physics, one could argue that the correct components of $d^2\mathbf{P}'/dt^2$ could be obtained on the sole basis of a geometrical argument. This is so.

That this point does not obviate the need for a general discussion of curvilinear coordinates, and, in particular, of differentiation in a curvilinear system, follows from the fact that our example is so simple, involving only a vector (first-rank tensor) and a simple motion. When changes in a quantity such as a second-rank tensor (stress being an important example) are considered, we can no longer lean on simple geometrical arguments to insure physically meaningful formulations.

Having suggested that life is not simple when one wishes to discuss tensors in curvilinear coordinates, we turn to a general discussion of this topic.

A. General Coordinate Transformations

We introduce first the concept of a coordinate transformation. We assume the existence of some coordinate system in which the position of a point is specified by giving the coordinates x^1, x^2, and x^3. (Superscripts, rather than subscripts, are used for a reason which will soon be clear. To avoid confusion the power of a coordinate will be denoted by $(x)^n$ rather than x^n.) Now suppose a new set of functions $\bar{x}^k$ could be defined by three equations

$$\bar{x}^k = \bar{x}^k(x^1, x^2, x^3) \quad \text{for } k = 1, 2, 3 \quad (A.67)$$

If these equations possess a unique inverse, that is, if also

$$x^i = x^i(\bar{x}^1, \bar{x}^2, \bar{x}^3) \quad \text{for } i = 1, 2, 3 \quad (A.68)$$

can be written, then the numbers $\bar{x}^k$ are the coordinates in some other system, and Eq. A.67 defines a *coordinate transformation*.

As an immediate example consider plane polar coordinates: We find

$$\begin{aligned} r &= r(x, y) = \sqrt{(x)^2 + (y)^2} \\ \theta &= \theta(x, y) = \tan^{-1}(y/x) \end{aligned} \quad (A.69)$$

and the inverse transformation

$$\begin{aligned} x &= x(r, \theta) = r \cos \theta \\ y &= y(r, \theta) = r \sin \theta \end{aligned}$$

as the sets corresponding to Eqs. A.67 and A.68.

In cartesian coordinates one usually defines the *base vectors* $\mathbf{i}_1$, $\mathbf{i}_2$, $\mathbf{i}_3$, in such a way that any cartesian vector may be expressed as $u = u^1 \mathbf{i}_1 + u^2 \mathbf{i}_2 + u^3 \mathbf{i}_3$. With the convention that we *sum over diagonally repeated indices*, this can be written $u = u^k \mathbf{i}_k$.

If the numbers z^m denote the *cartesian* components of a point, then the curvilinear base vectors $\mathbf{g}_k$, of a system with coordinates x^k, may be defined by

$$\mathbf{g}_k = \frac{\partial z^m}{\partial x^k} \mathbf{i}_m \quad (A.70)*$$

Thus, again using polar coordinates as an example, we find

$$\mathbf{g}_r = \frac{\partial x}{\partial r} \mathbf{i}_1 + \frac{\partial y}{\partial r} \mathbf{i}_2 = \cos \theta \mathbf{i}_1 + \sin \theta \mathbf{i}_2$$

$$\mathbf{g}_\theta = \frac{\partial x}{\partial \theta} \mathbf{i}_1 + \frac{\partial y}{\partial \theta} \mathbf{i}_2 = -r \sin \theta \mathbf{i}_1 + r \cos \theta \mathbf{i}_2$$

---

* The cartesian base vectors are by definition unit vectors, while the curvilinear base vectors are generally not unit vectors.

Now let us consider length in a curvilinear system. If, for example, **u** is a position vector in cartesian coordinates, then a differential line element is given by
$$d\mathbf{u} = du^k \mathbf{i}_k \tag{A.71}$$
The length, or magnitude, of $d\mathbf{u}$ is given by the square root of
$$(ds)^2 = d\mathbf{u} \cdot d\mathbf{u} = (du^k \mathbf{i}_k) \cdot (du^m \mathbf{i}_m) \tag{A.72}$$
(The dummy index has been changed in one term of Eq. A.72 to avoid ambiguity.)

Now introducing a curvilinear system x^n, we have, by the chain rule of differentiation,
$$du^k = \frac{\partial u^k}{\partial x^n} dx^n \tag{A.73}$$
Hence Eq. A.72 becomes
$$(ds)^2 = \left(\frac{\partial u^k}{\partial x^n} \mathbf{i}_k\right) \cdot \left(\frac{\partial u^m}{\partial x^p} \mathbf{i}_m\right)(dx^n \, dx^p) \tag{A.74}$$
But, since u^k are cartesian components, Eq. A.70 may be used here, and
$$(ds)^2 = \mathbf{g}_n \cdot \mathbf{g}_p \, dx^n \, dx^p \tag{A.75}$$
We now define
$$\mathbf{g}_n \cdot \mathbf{g}_p = g_{np} \tag{A.76}$$
to be the *fundamental metric tensor*, and
$$(ds)^2 = g_{np} dx^n \, dx^p \tag{A.77}$$
shows that g_{np} allows us to find the distance between neighboring points in a curvilinear system. It is clear from Eq. A.77 that g_{np} is symmetric in its indices.

In polar coordinates:
$$g_{11} = \mathbf{g}_r \cdot \mathbf{g}_r = 1$$
$$g_{22} = \mathbf{g}_\theta \cdot \mathbf{g}_\theta = r^2 \quad \text{or} \quad g_{ij} = \begin{pmatrix} 1 & 0 & 0 \\ 0 & r^2 & 0 \\ 0 & 0 & 0 \end{pmatrix}$$
$$g_{12} = g_{21} = \mathbf{g}_r \cdot \mathbf{g}_\theta = 0$$

Hence we obtain the familiar result
$$(ds)^2 = (dr)^2 + (r \, d\theta)^2$$

If the curvilinear coordinates are orthogonal, then the base vectors are mutually perpendicular, and
$$\mathbf{g}_p \cdot \mathbf{g}_n = g_{pn} = 0 \quad \text{if} \quad p \neq n \tag{A.78}$$
Hence g_{pn} is a diagonal tensor in orthogonal systems.

It will prove useful in some later applications to have a tensor g^{pm} (note superscripts) defined in such a way that

$$g_{pn}g^{pm} = \delta_n^m \tag{A.79}$$

where δ_n^m is a Kronecker delta. The tensor g^{pm} is said to be *conjugate* to g_{pn}.

In polar coordinates we find

$$g_{11}g^{11} = 1 = g^{11}$$
$$g_{22}g^{22} = 1 = r^2 g^{22}$$

so that

$$g^{ij} = \begin{pmatrix} 1 & 0 & 0 \\ 0 & 1/r^2 & 0 \\ 0 & 0 & 0 \end{pmatrix}$$

By now the reader may be sufficiently annoyed at this arbitrary use of subscripts and superscripts to welcome the establishment of some further definitions.

B. Contravariant and Covariant Tensors

If three numbers u^m, referred to a system of coordinates x^m, are related to three numbers $\bar{u}^k$, referred to coordinates $\bar{x}^k$, by

$$\bar{u}^k = \frac{\partial \bar{x}^k}{\partial x^m} u^m \tag{A.80}$$

then the quantities $\bar{u}^k$ and u^m are called the *contravariant components* of a vector.

For an example of a contravariant vector consider the functional relationship given in Eq. A.67. By the chain rule

$$d\bar{x}^k = \frac{\partial \bar{x}^k}{\partial x^m} dx^m \tag{A.81}$$

Hence differentials of position are components of a contravariant vector. By convention we shall use a superscript to denote contravariance.

In a similar manner, if two sets of numbers are related by

$$u_m = \frac{\partial \bar{x}^k}{\partial x^m} \bar{u}_k \tag{A.82}$$

they are said to be components of a *covariant* vector. The partial derivatives of a scalar form such a vector, for, by the chain rule

$$\frac{\partial \phi}{\partial x^m} = \frac{\partial \phi}{\partial \bar{x}^k} \frac{\partial \bar{x}^k}{\partial x^m} \tag{A.83}$$

By convention we use a subscript to denote covariance. Furthermore, if we regard a superscript in the denominator, as in Eq. A.83, as a subscript in the numerator, then $\partial \phi / \partial \bar{x}^k$ is covariant, consistent with this convention.

We can define three types of second-order tensors, depending upon their rules of transformation:

$$\bar{T}^{km} = \frac{\partial \bar{x}^k}{\partial x^i} \frac{\partial \bar{x}^m}{\partial x^j} T^{ij} \quad \text{(contravariant tensor)} \tag{A.84}$$

$$\bar{T}_{km} = \frac{\partial x^i}{\partial \bar{x}^k} \frac{\partial x^j}{\partial \bar{x}^m} T_{ij} \quad \text{(covariant tensor)} \tag{A.85}$$

$$\bar{T}^k_m = \frac{\partial \bar{x}^k}{\partial x^i} \frac{\partial x^j}{\partial \bar{x}^m} T^i_j \quad \text{(mixed tensor)} \tag{A.86}$$

We should remind the reader that we are free to *define* anything we please. Whether any physical entities behave like the components of, say, a contravariant tensor is another question, and one to which we must later direct some attention.

Let us consider immediately the special case of a coordinate transformation between two cartesian systems. Then Eqs. A.15 and A.16, with a modified notation, become

$$x^i = a_{ik} \bar{x}^k \tag{A.87}$$

$$\bar{x}^k = a_{ik} x^i \tag{A.88}$$

At the moment we do not know the tensorial character of a_{ik}, so we leave the notation unchanged. It follows immediately that

$$\frac{\partial x^i}{\partial \bar{x}^k} = a_{ik} = \frac{\partial \bar{x}^k}{\partial x^i} \tag{A.89}$$

By inspection of the definitions of covariance and contravariance, it should be obvious that for cartesian tensors

$$T^{km} = T_{km} = T^k_m \tag{A.90}$$

In short, as long as one is restricted to cartesian tensors, he need not be concerned with covariance and contravariance.

Before examining in detail the manipulation of non-cartesian tensors, we establish some simple conventions and rules. We have already established the convention that summation is taken over *diagonally* repeated indices, so that, for example

$$\bar{u}_k = \frac{\partial x^m}{\partial \bar{x}^k} u_m = \frac{\partial x^1}{\partial \bar{x}^k} u_1 + \frac{\partial x^2}{\partial \bar{x}^k} u_2 + \frac{\partial x^3}{\partial \bar{x}^k} u_3$$

It is not difficult to show that the addition of two tensors of the same order and type produces a new tensor of that order and type:

$$C_{ij}^k = A_{ij}^k + B_{ij}^k$$

A number of quotient rules can be proved which can be used to establish the tensorial character of an entity. For example, if u^k is a contravariant vector, and

$$C = v_k u^k$$

is an invariant, then v_k must be a covariant vector. It might be useful to illustrate the proof. Note that although v_k is written with a subscript, in anticipation of the result, we must treat this strictly as a notation. For example, we do not know whether summation is allowed over k.

Since C is invariant, then in another coordinate system

$$\bar{C} = C = \bar{v}_k \bar{u}^k$$

By definition

$$\bar{u}^k = \frac{\partial \bar{x}^k}{\partial x^m} u^m$$

so that

$$C = \bar{v}_k \frac{\partial \bar{x}^k}{\partial x^m} u^m$$

But, since k is a dummy index (whether we sum over it or not) the definition of C can be rewritten as $C = v_k u^k = v_m u^m$ and it follows that

$$\left(v_m - \bar{v}_k \frac{\partial \bar{x}^k}{\partial x^m} \right) u^m = 0$$

Since the vector u^m has not been specified, we can consider it to be non-zero. As a result we obtain

$$v_m = \bar{v}_k \frac{\partial \bar{x}^k}{\partial x^m}$$

and v_m is shown to be a covariant vector by comparison with the definition A.82.

Another quotient rule is the following:

If v^k is an arbitrary contravariant vector and u_m is a covariant vector, and if

$$u_m = G_{mk} v^k$$

then G_{mk} is a covariant second-rank tensor.

As a consequence of these two quotient rules we may prove that the fundamental metric tensor g_{np} (Eq. A.77) is a second-rank covariant tensor. From A.77 we have

$$(ds)^2 = g_{np} dx^n dx^p$$

Since $(ds)^2$ is the distance between two points, it is an invariant. By the first quotient rule above, since dx^p is a contravariant vector, the term $g_{np}\,dx^n$ must be a covariant vector. By the second quotient rule, since dx^n is a contravariant vector, g_{np} must be a second-rank covariant tensor. Hence our assertion is established.

As a corollary to this quotient rule it follows that $G_{ij}u^j$ is a covariant vector. In the special case that the tensor G_{ij} is the fundamental metric tensor, the operation

$$u_i = g_{ij}u^j \tag{A.91}$$

defines the *associate vector* of u^j.

In the same way, if we define

$$v^i = g^{ij}v_j \tag{A.92}$$

then v^i is the associate vector of v_j. These two operations are usually referred to as "lowering (or raising) the index."

C. Christoffel Symbols

In the course of differentiating tensors in curvilinear systems we find that certain combinations of derivatives of the fundamental tensor g_{ij} occur repeatedly. For the sake of conciseness, and because they are of central importance, we honor them with special symbols.

The *Christoffel symbols* of the first and second kinds are defined, respectively, by

$$[ij, k] = \frac{1}{2}\left(\frac{\partial g_{ik}}{\partial x^j} + \frac{\partial g_{jk}}{\partial x^i} - \frac{\partial g_{ij}}{\partial x^k}\right) \tag{A.93}$$

and

$$\left\{\begin{matrix}m\\ij\end{matrix}\right\} = g^{mk}[ij, k] \tag{A.94}$$

From Eq. A.93 it is obvious that

$$[ij, k] = [ji, k] \tag{A.95}$$

We note also that, by definition

$$[kj, i] = \frac{1}{2}\left(\frac{\partial g_{ki}}{\partial x^j} + \frac{\partial g_{ji}}{\partial x^k} - \frac{\partial g_{kj}}{\partial x^i}\right) \tag{A.96}$$

But we have already shown $g_{ij} = g_{ji}$, so that

$$[kj, i] = \frac{1}{2}\left(\frac{\partial g_{ik}}{\partial x^j} + \frac{\partial g_{ij}}{\partial x^k} - \frac{\partial g_{jk}}{\partial x^i}\right) \tag{A.97}$$

Adding Eqs. A.97 and A.93 we find

$$\frac{\partial g_{ik}}{\partial x^j} = [ij, k] + [kj, i] \tag{A.98}$$

We shall have need for the corresponding expression for the conjugate to g_{ik}. From Eq. A.79

$$g_{pn}g^{pm} = \delta_n^m$$

$$\frac{\partial \delta_n^m}{\partial x^j} = 0 = g_{pn}\frac{\partial g^{pm}}{\partial x^j} + g^{pm}\frac{\partial g_{pn}}{\partial x^j}$$

Multiply by g^{nk}

$$0 = g_{pn}g^{nk}\frac{\partial g^{pm}}{\partial x^j} + g^{nk}g^{pm}\frac{\partial g_{pn}}{\partial x^j}$$

But

$$g_{pn}g^{nk} = \delta_p^k$$

so

$$0 = \frac{\partial g^{km}}{\partial x^j} + g^{nk}g^{pm}\frac{\partial g_{pn}}{\partial x^j}$$

Solving for $\partial g^{km}/\partial x^j$ and using Eq. A.98, we find

$$\frac{\partial g^{km}}{\partial x^j} = -g^{nk}g^{pm}([pj, n] + [nj, p])$$

$$= -g^{pm}\begin{Bmatrix}k\\pj\end{Bmatrix} - g^{nk}\begin{Bmatrix}m\\nj\end{Bmatrix} \tag{A.99}$$

Equations A.98 and A.99 will be useful in subsequent developments.

We now examine the transformation laws of the Christoffel symbols. Since the fundamental tensor is covariant, it transforms according to

$$\bar{g}_{km} = \frac{\partial x^i}{\partial \bar{x}^k}\frac{\partial x^j}{\partial \bar{x}^m}g_{ij} \tag{A.100}$$

If this is differentiated with respect to $\bar{x}^n$ we find

$$\frac{\partial \bar{g}_{km}}{\partial \bar{x}^n} = \frac{\partial x^i}{\partial \bar{x}^k}\frac{\partial x^j}{\partial \bar{x}^m}\frac{\partial x^p}{\partial \bar{x}^n}\frac{\partial g_{ij}}{\partial x^p} + \frac{\partial^2 x^i}{\partial \bar{x}^k \partial \bar{x}^n}\frac{\partial x^j}{\partial \bar{x}^m}g_{ij} + \frac{\partial x^i}{\partial \bar{x}^k}\frac{\partial^2 x^j}{\partial \bar{x}^m \partial \bar{x}^n}g_{ij} \tag{A.101}$$

Two similar equations can be obtained, by inspection, upon cyclic interchange of the indices k, m, and n. If Eq. A.101 is subtracted from the sum of these two equations, and the result divided by two, we obtain

$$\overline{[km, n]} = [ij, p]\frac{\partial x^i}{\partial \bar{x}^k}\frac{\partial x^j}{\partial \bar{x}^m}\frac{\partial x^p}{\partial \bar{x}^n} + g_{ij}\frac{\partial x^i}{\partial \bar{x}^n}\frac{\partial^2 x^j}{\partial \bar{x}^k \partial \bar{x}^m} \tag{A.102}$$

The transformation of the conjugate tensor g^{nr} is

$$\bar{g}^{nr} = g^{st} \frac{\partial \bar{x}^n}{\partial x^s} \frac{\partial \bar{x}^r}{\partial x^t} \tag{A.103}$$

If the left- and right-hand sides of A.102 are multiplied by the corresponding sides of A.103, the result is

$$\overline{\left\{ \begin{array}{c} r \\ km \end{array} \right\}} = \left\{ \begin{array}{c} t \\ ij \end{array} \right\} \frac{\partial \bar{x}^r}{\partial x^t} \frac{\partial x^i}{\partial \bar{x}^k} \frac{\partial x^j}{\partial \bar{x}^m} + \frac{\partial \bar{x}^r}{\partial x^j} \frac{\partial^2 x^j}{\partial \bar{x}^k \partial \bar{x}^m} \tag{A.104}$$

Equations A.102 and A.104 show that neither of the Christoffel symbols defines a tensor, except in the event of a linear transformation, for which $\partial^2 x^j / \partial \bar{x}^k \partial \bar{x}^m = 0$.

We require one more relationship before we are prepared to consider the important topic of differentiation. If both sides of Eq. A.104 are multiplied by $\partial x^s / \partial \bar{x}^r$, the result is

$$\frac{\partial^2 x^s}{\partial \bar{x}^k \partial \bar{x}^m} = \overline{\left\{ \begin{array}{c} r \\ km \end{array} \right\}} \frac{\partial x^s}{\partial \bar{x}^r} - \left\{ \begin{array}{c} s \\ ij \end{array} \right\} \frac{\partial x^i}{\partial \bar{x}^k} \frac{\partial x^j}{\partial \bar{x}^m} \tag{A.105}$$

Hence the second derivatives are expressible in terms of the first derivatives and the Christoffel symbols.

As an illustration of these results, let us consider a transformation such that the x^k are cartesian components and the $\bar{x}^k$ are plane polar components. We have already found that

$$\bar{g}_{ij} = \begin{pmatrix} 1 & 0 & 0 \\ 0 & r^2 & 0 \\ 0 & 0 & 0 \end{pmatrix}$$

In cartesian coordinates we know that

$$(ds)^2 = \delta_{ij} \, dx^i \, dx^j = (dx^1)^2 + (dx^2)^2 + (dx^3)^2$$

Hence, by comparison with Eq. A.77, we see that

$$g_{ij} = \delta_{ij} = \begin{pmatrix} 1 & 0 & 0 \\ 0 & 1 & 0 \\ 0 & 0 & 1 \end{pmatrix}$$

or, in words, the Kronecker delta is the fundamental metric tensor of cartesian coordinates. Since all components of δ are constant, it follows immediately from Eqs. A.93 and A.94 that

$$[ij, k] = \left\{ \begin{array}{c} m \\ ij \end{array} \right\} = 0$$

in cartesian coordinates.

For polar coordinates we find

$$-\overline{[22, 1]} = \overline{[12, 2]} = \overline{[21, 2]} = r$$

$$\overline{\left\{\begin{matrix}1\\22\end{matrix}\right\}} = -r \quad \overline{\left\{\begin{matrix}2\\12\end{matrix}\right\}} = \overline{\left\{\begin{matrix}2\\21\end{matrix}\right\}} = \frac{1}{r}$$

All other Christoffel symbols are zero in polar coordinates.

D. Covariant Differentiation

Now let us consider the contravariant vector

$$u^k = \frac{\partial x^k}{\partial \bar{x}^i} \bar{u}^i$$

We shall be interested in its derivative

$$\frac{\partial u^k}{\partial x^j} = \frac{\partial \bar{u}^i}{\partial \bar{x}^n} \frac{\partial \bar{x}^n}{\partial x^j} \frac{\partial x^k}{\partial \bar{x}^i} + \bar{u}^i \frac{\partial^2 x^k}{\partial \bar{x}^i \partial \bar{x}^n} \frac{\partial \bar{x}^n}{\partial x^j} \tag{A.106}$$

Comparison of Eqs. A.106 and A.86 shows that $\partial u^k/\partial x^j$ would be a mixed tensor but for the presence of the second term in A.106. Now Eq. A.105 is used to eliminate the second derivative term, and A.106 becomes

$$\frac{\partial u^k}{\partial x^j} = \frac{\partial \bar{u}^i}{\partial \bar{x}^n} \frac{\partial \bar{x}^n}{\partial x^j} \frac{\partial x^k}{\partial \bar{x}^i} + \bar{u}^i \frac{\partial \bar{x}^n}{\partial x^j} \left[\overline{\left\{\begin{matrix}p\\in\end{matrix}\right\}} \frac{\partial x^k}{\partial \bar{x}^p} - \left\{\begin{matrix}k\\rs\end{matrix}\right\} \frac{\partial x^r}{\partial \bar{x}^i} \frac{\partial x^s}{\partial \bar{x}^n} \right]$$

After some manipulation, this can be put in the form

$$\frac{\partial u^k}{\partial x^j} + \left\{\begin{matrix}k\\rj\end{matrix}\right\} u^r = \frac{\partial \bar{x}^n}{\partial x^j} \frac{\partial x^k}{\partial \bar{x}^i} \left[\frac{\partial \bar{u}^i}{\partial \bar{x}^n} + \overline{\left\{\begin{matrix}i\\rn\end{matrix}\right\}} \bar{u}^r \right] \tag{A.107}$$

If A.107 is compared to A.86, it would appear that the sum

$$\frac{\partial u^k}{\partial x^j} + \left\{\begin{matrix}k\\rj\end{matrix}\right\} u^r$$

is a mixed tensor of the second rank. It is called the *covariant derivative* of u^k with respect to x^j, and is usually denoted by the so-called "comma notation":

$$u^k_{,j} = \frac{\partial u^k}{\partial x^j} + \left\{\begin{matrix}k\\rj\end{matrix}\right\} u^r \tag{A.108}$$

In a similar manner the *covariant derivative* of a *covariant* vector may be defined by

$$u_{k,j} = \frac{\partial u_k}{\partial x^j} - \left\{\begin{matrix}r\\kj\end{matrix}\right\} u_r \tag{A.109}$$

and $u_{k,j}$ may be shown to be a covariant tensor.

We can now define three vector derivatives of central importance in a discussion of physics. The *divergence* of a contravariant vector is defined as

$$\text{div } u^k = u^k_{,k} = \frac{\partial u^k}{\partial x^k} + \begin{Bmatrix} k \\ ik \end{Bmatrix} u^i \qquad (A.110)$$

The divergence of a covariant vector is defined as

$$\text{div } u_k = g^{jk} u_{j,k} \qquad (A.111)$$

The *gradient* of a scalar, in curvilinear coordinates, is defined as a covariant vector with components

$$\text{grad } \phi = \frac{\partial \phi}{\partial x^k} \qquad (A.112)$$

and requires no modification over its definition in cartesian coordinates. Since grad ϕ is a covariant vector, its divergence is

$$\text{div grad } \phi = g^{jk} \left(\frac{\partial \phi}{\partial x^j} \right)_{,k} = g^{jk} \left[\frac{\partial}{\partial x^k} \left(\frac{\partial \phi}{\partial x^j} \right) - \begin{Bmatrix} r \\ jk \end{Bmatrix} \frac{\partial \phi}{\partial x^r} \right]$$

This operation defines the Laplacian of ϕ, and is written as

$$\nabla^2 \phi = g^{jk} \left(\frac{\partial^2 \phi}{\partial x^k \partial x^j} - \begin{Bmatrix} r \\ jk \end{Bmatrix} \frac{\partial \phi}{\partial x^r} \right) \qquad (A.113)$$

In plane cartesian coordinates

$$\text{div } u_k = \text{div } u^k = \frac{\partial u_1}{\partial x_1} + \frac{\partial u_2}{\partial x_2}$$

$$\text{grad } \phi = (\partial \phi / \partial x_1, \partial \phi / \partial x_2)$$

$$\nabla^2 \phi = \frac{\partial^2 \phi}{\partial x_1^2} + \frac{\partial^2 \phi}{\partial x_2^2}$$

In plane polar coordinates

$$\text{div } u^k = \frac{\partial u^1}{\partial r} + \frac{\partial u^2}{\partial \theta} + \frac{u^1}{r}$$

$$\text{grad } \phi = (\partial \phi / \partial r, \partial \phi / \partial \theta)$$

$$\nabla^2 \phi = \frac{\partial^2 \phi}{\partial r^2} + \frac{1}{r^2} \left[\frac{\partial^2 \phi}{\partial \theta^2} + r \frac{\partial \phi}{\partial r} \right] = \frac{\partial^2 \phi}{\partial r^2} + \frac{1}{r} \frac{\partial \phi}{\partial r} + \frac{1}{r^2} \frac{\partial^2 \phi}{\partial \theta^2}$$

The covariant derivatives of second-order tensors may be obtained by methods analogous to, but more tedious than, those illustrated for vectors.

The results are

$$A^{kp}_{,m} = \frac{\partial A^{kp}}{\partial x^m} + \begin{Bmatrix} k \\ mn \end{Bmatrix} A^{np} + \begin{Bmatrix} p \\ mn \end{Bmatrix} A^{kn} \quad \text{(A.114)}$$

$$A^{k}_{p,m} = \frac{\partial A^{k}_{p}}{\partial x^m} - \begin{Bmatrix} n \\ pm \end{Bmatrix} A^{k}_{n} + \begin{Bmatrix} k \\ mn \end{Bmatrix} A^{n}_{p} \quad \text{(A.115)}$$

$$A_{kp,m} = \frac{\partial A_{kp}}{\partial x^m} - \begin{Bmatrix} n \\ km \end{Bmatrix} A_{np} - \begin{Bmatrix} n \\ pm \end{Bmatrix} A_{kn} \quad \text{(A.116)}$$

Finally we consider a contravariant vector u^k which is a function of some parameter t along a curve $x^m = x^m(t)$. It is possible to define a contravariant vector, known as the *intrinsic derivative* of u^k, whose components are

$$\frac{\delta u^k}{\delta t} = u^k_{,m} \frac{dx^m}{dt} = \left[\frac{\partial u^k}{\partial x^m} + \begin{Bmatrix} k \\ rm \end{Bmatrix} u^r \right] \frac{dx^m}{dt} \quad \text{(A.117)}$$

Since

$$\frac{\partial u^k}{\partial x^m} \frac{dx^m}{dt} = \frac{du^k}{dt}$$

the intrinsic derivative may be written as

$$\frac{\delta u^k}{\delta t} = \frac{du^k}{dt} + \begin{Bmatrix} k \\ rm \end{Bmatrix} u^r \frac{dx^m}{dt} \quad \text{(A.118)}$$

As we have already stated, we are free to *define* nearly anything we please. But if this effort is to be more than mathematical exercise we must show some correspondence between the physical quantities we wish to describe and the mathematical quantities we have defined. Let us return, then, to the problem which initiated this chapter, and discuss the acceleration of a particle traveling along a path $x^m = x^m(t)$.

We *define* the *velocity* to be the contravariant vector

$$v^m = dx^m/dt$$

We *define* the *acceleration* to be the contravariant vector

$$a^m = \delta v^m/\delta t$$

In cartesian coordinates, the Christoffel symbols vanish, and

$$a^m = \frac{\delta v^m}{\delta t} = \frac{dv^m}{dt} = \frac{d^2 x^m}{dt^2}$$

In plane polar coordinates, however, we find

$$a^1 = \frac{dv^1}{dt} - rv^2\frac{dx^2}{dt} = \frac{dv^1}{dt} - r(v^2)^2$$

$$a^2 = \frac{dv^2}{dt} + \frac{1}{r}\left(v^2\frac{dx^1}{dt} + v^1\frac{dx^2}{dt}\right) = \frac{dv^2}{dt} + \frac{2}{r}v^1v^2$$

If the path is a circle of radius R, $v^1 = 0$ and if $v^2 = \omega$ = constant, we find

$$a^1 = -R(\omega)^2$$
$$a^2 = 0$$

as before. On the basis of this isolated example we would conclude tentatively (but correctly) that the acceleration of a particle, referred to curvilinear coordinates, is the intrinsic derivative of its velocity. The single formula of Eq. A.118 then suffices to describe acceleration in any system, without the introduction of artificialities, such as "centrifugal acceleration," and without the appeal to (often messy) geometrical arguments.

E. Physical Components

In general, each coordinate of a curvilinear system is not of the same dimension. In polar coordinates, for example, the radial coordinate has the dimension of length, but the angular coordinate is dimensionless. As a result, the angular velocity $d\theta/dt$ does not have the dimensions of a velocity. We now examine the possibility of redefining the components of tensors in such a way that all components have the same dimensions.

We begin by considering the contravariant vector u^k. The "length" of u^k will be defined as

$$|u^k| = (g_{ij}u^i u^j)^{1/2} \tag{A.119}$$

If we restrict our attention to orthogonal coordinate systems (not a very serious restriction), then

$$|u^k|^2 = g_{11}(u^1)^2 + g_{22}(u^2)^2 + g_{33}(u^3)^2 \tag{A.120}$$

It is common to introduce a set of *scale factors*, defined so that

$$(h_1)^2 = g_{11}$$
$$(h_2)^2 = g_{22} \tag{A.121}$$
$$(h_3)^2 = g_{33}$$

Then

$$|u^k|^2 = (h_1 u^1)^2 + (h_2 u^2)^2 + (h_3 u^3)^2 \tag{A.122}$$

Since Eq. A.122 must be dimensionally homogeneous, it must be true

TENSORS 227

that each term of the form $h_j u^j$ is of the same dimensions as the others. On this basis we define the *physical components* of the vector u^k to be

$$u^{(j)} = h_j u^j \qquad \text{(no summation on } j\text{)} \tag{A.123}$$

The parenthetical notation is used to distinguish the physical components, which do not transform as tensors, from the contravariant components.

Since

$$g_{ij} = \begin{pmatrix} 1 & 0 & 0 \\ 0 & r^2 & 0 \\ 0 & 0 & 0 \end{pmatrix}$$

for plane polar coordinates, the scale factors are $h_1 = 1$, $h_2 = r$, $h_3 = 0$.

The physical components of velocity, in polar coordinates, are

$$\begin{array}{cc} u^{(1)} = u^1 & u^{(r)} = u^r \\ u^{(2)} = ru^2 & u^{(\theta)} = ru^\theta \end{array} \quad \text{or} \quad$$

The transformation law of physical components can be readily deduced. Since

$$\bar{u}^i = \frac{\partial \bar{x}^i}{\partial x^j} u^j$$

we find

$$\frac{\bar{u}^{(i)}}{\bar{h}_i} = \frac{\partial \bar{x}^i}{\partial x^j} \frac{u^{(j)}}{h_j}$$

or

$$\bar{u}^{(i)} = \frac{\bar{h}_i}{h_j} \frac{\partial \bar{x}^i}{\partial x^j} u^{(j)} \tag{A.124}$$

In Eq. A.124 we sum on j but not i, so that

$$\bar{u}^{(i)} = \bar{h}_i \left[\frac{1}{h_1} \frac{\partial \bar{x}^i}{\partial x^1} u^{(1)} + \frac{1}{h_2} \frac{\partial \bar{x}^i}{\partial x^2} u^{(2)} + \frac{1}{h_3} \frac{\partial \bar{x}^i}{\partial x^3} u^{(3)} \right]$$

In an analogous manner the physical components of a covariant vector may be defined by

$$u_{(j)} = h^j u_j \qquad \text{(no summation on } j\text{)} \tag{A.125}$$

where the scale factors h^j are defined in terms of the conjugate metric g^{ij}:

$$(h^1)^2 = g^{11} \quad \text{etc.}$$

A very important covariant vector is the gradient of a scalar, defined in Eq. A.112 as

$$\text{grad } \phi = \frac{\partial \phi}{\partial x^k} = \nabla_k \phi$$

where the notation ∇_k is taken to indicate covariance. Hence the physical components of the gradient are

$$\nabla_{(k)}\phi = h^k \nabla_k \phi \quad \text{(no summation on } k\text{)} \tag{A.126}$$

In plane polar coordinates the components of the gradient are

$$\nabla_k \phi = \left(\frac{\partial \phi}{\partial r}, \frac{\partial \phi}{\partial \theta} \right)$$

Its physical components are

$$\nabla_{(k)}\phi = \left(\frac{\partial \phi}{\partial r}, \frac{1}{r} \frac{\partial \phi}{\partial \theta} \right)$$

We are often concerned with equations involving second-order tensors of the form

$$v^i = T^{ij} u_j \tag{A.127}$$

$$v^i = T^i_j u^j \tag{A.128}$$

$$v_i = T_{ij} u^j \tag{A.129}$$

The physical components of a second-order tensor will be defined in such a way that, for example

$$v^{(i)} = T^{(ij)} u_{(j)} \tag{A.130}$$

From Eqs. A.127, and A.123 and A.125 we find

$$\frac{v^{(i)}}{h_i} = T^{ij} \frac{u_{(j)}}{h^j} \tag{A.131}$$

Comparing Eq. A.131 to A.130 we obtain

$$T^{(ij)} = \frac{h_i}{h^j} T^{ij} = h_i h_j T^{ij} \tag{A.132}$$

As a final example, let us calculate the physical components of the *rate of deformation tensor*, which we define as

$$\Delta^i_j = (u^i_{,j} + u^j_{,i})$$

where u^i is the contravariant velocity vector, and the comma notation of A.108 holds.

The physical components of Δ^i_j are defined to be

$$\Delta^{(i)}_{(j)} = (u^{(i)}_{,(j)} + u^{(j)}_{,(i)})$$

Hence

$$\Delta^{(i)}_{(j)} = \left[\frac{h_i}{h_j} u^i_{,j} + \frac{h_j}{h_i} u^j_{,i} \right]$$

For example

$$\Delta^{(1)}_{(2)} = \Delta^{(r)}_{(\theta)} = \left[\frac{h_1}{h_2}\left(\frac{\partial u^1}{\partial x^2} + \left\{{1 \atop r2}\right\}u^r\right) + \frac{h_2}{h_1}\left(\frac{\partial u^2}{\partial x^1} + \left\{{2 \atop r1}\right\}u^r\right)\right]$$

Introducing the known values for the scale factors and the Christoffel symbols, we obtain, with $x^1 = r$ and $x^2 = \theta$

$$\Delta^{(r)}_{(\theta)} = \frac{1}{r}\frac{\partial u^r}{\partial \theta} - u^\theta + r\frac{\partial u^\theta}{\partial r} + u^\theta = \frac{1}{r}\frac{\partial u^r}{\partial \theta} - u^\theta + \frac{\partial}{\partial r}(ru^\theta)$$

We must now introduce the physical components of velocity. From Eq. A.123 and the example following, we found

$$u^{(r)} = u^r$$
$$u^{(\theta)} = ru^\theta$$

Our final result, then, is

$$\Delta^{(r)}_{(\theta)} = \left[\frac{1}{r}\frac{\partial u^{(r)}}{\partial \theta} - \frac{u^{(\theta)}}{r} + \frac{\partial u^{(\theta)}}{\partial r}\right]$$

This result is commonly presented in texts on hydrodynamics, although it is rare that reference is made to the fact that physical components are presented, rather than tensorial components.

Standard texts, from which much of the material of this section has been drawn, include the following books:

A1. Aris, R. *Vectors, Tensors, and the Basic Equations of Fluid Mechanics*, Prentice-Hall, Englewood Cliffs, N. J., 1962.
A2. Jeffreys, H., *Cartesian Tensors*, Cambridge University Press, Cambridge, 1961.
A3. McConnell, A. J., *Applications of Tensor Analysis*, Dover, New York, 1957.
A4. Spain, B., *Tensor Calculus*, Interscience, New York, 1953.
A5. Temple, G., *Cartesian Tensors*, Methuen, London, 1960.

Appendix B

Kinematics

Kinematics is the study of motion, with no regard for the forces involved in that motion. A complete kinematical description of material response would allow one to give the spatial coordinates of every material point at any instant of time, given some set of initial conditions. We wish to investigate in this section the mathematics of response per se, without regard for the relation of response to forces.

I. MATERIAL VS. SPATIAL COORDINATES

Rheology is naturally concerned with motion. While motion is an intuitive concept, its proper mathematical description, particularly in rheological equations of state, is not at all simple. We shall first develop some general concepts of motion before turning to a discussion of particular types of motion of importance in rheology.

Let us consider a material in motion with respect to some fixed cartesian coordinate system. At some instant of time t_0, the positions of the *particles* of the material may be labeled with coordinates $\mathbf{X}$. Here the term particle is used only in the sense that if any point in space is occupied by material, that place corresponds to a particle of material. We imagine that we can somehow "tag" particles, and thereby locate any particle at some time as one with a given initial position $\mathbf{X}$. At some later time t, the particles that were at $\mathbf{X}$ now occupy the places $\mathbf{x}$. A *motion* is defined by a transformation of particles through places in the course of time. Mathematically this is stated by

$$\mathbf{X} = \mathbf{X}(\mathbf{x}, t) \tag{B.1}$$

or its inverse

$$\mathbf{x} = \mathbf{x}(\mathbf{X}, t) \tag{B.2}$$

Equation B.1 states that particles in motion occupy different places as time passes. Its inverse states that places in space, through which a material is in motion, will be occupied by different particles as the motion proceeds in time. It is the essence of motion that material moves through space

KINEMATICS

during the course of time. Furthermore, if the inverse transformation is assumed unique, we have stated the principle of impenetrability of matter: if two particles are distinct at some time, they will always be distinct.

The variables **X** are spoken of as *material coordinates*, because they are labels attached to particles of material distinguished by their initial positions, while the variables **x** are called *spatial coordinates*. It is sometimes more appropriate to work with a material description of behavior and sometimes with a spatial description. For example, spatial, or fixed, coordinates are usually the natural coordinate system of measurement, since most measuring instruments are held fixed in space. Thus spatial coordinates are often referred to as "the laboratory coordinate system." On the other hand material response occurs as the material moves through space, and is inherently connected to a material coordinate system.

It should be clear that one can define two types of time derivatives: one keeping material coordinates constant $(\partial/\partial t)_\mathbf{X}$, and one keeping spatial coordinates constant $(\partial/\partial t)_\mathbf{x}$. As an example consider the time rate of change of the spatial coordinates of a particular particle of fluid. Since we confine our attention to a particular particle, we are keeping the material coordinates constant. Hence this derivative is

$$\left(\frac{\partial x_i}{\partial t}\right)_\mathbf{X} = v_i \tag{B.3}$$

and from this definition of v_i we see that we obtain just the velocity of a particle of material.

It is not difficult to establish a connection between the material and spatial derivatives of a quantity. Let us consider some scalar function of time and particle, say $f(\mathbf{X}, t)$. We seek the derivative of f following the motion, and hence for fixed **X**. From Eq. B.1 we may always express $f(\mathbf{X}, t)$ as $f[\mathbf{x}(\mathbf{X}, t), t]$. Application of the chain rules for differentiation gives

$$\left(\frac{\partial f}{\partial t}\right)_\mathbf{X} = \left(\frac{\partial}{\partial t} f[\mathbf{x}(\mathbf{X}, t), t]\right)_\mathbf{X} = \frac{\partial f}{\partial x_i}\left(\frac{\partial x_i}{\partial t}\right)_\mathbf{X} + \left(\frac{\partial f}{\partial t}\right)_\mathbf{x} \tag{B.4}$$

If Eq. B.3 is used this can be written as

$$\left(\frac{\partial f}{\partial t}\right)_\mathbf{X} = \left(\frac{\partial f}{\partial t}\right)_\mathbf{x} + v_i \frac{\partial f}{\partial x_i} \tag{B.5}$$

This derivative is usually called the "material derivative," or the "derivative following the motion." A common notation, which avoids the need to carry subscripts indicating the variables held constant in the differentiation, is

$$\left(\frac{\partial f}{\partial t}\right)_\mathbf{X} = \frac{Df}{Dt} = \frac{\partial f}{\partial t} + v_i \frac{\partial f}{\partial x_i} \tag{B.6}$$

In other words, the material derivative states that if one travels with the fluid, he sees apparent time changes which are due not only to unsteady state effects ($\partial f/\partial t$), but in addition variables appear to change with time because one is moving through a spatially inhomogeneous field ($v_i \, \partial f/\partial x_i$).

Recall that the function f we have considered is a scalar function, and hence these time derivatives must be (and, of course, are) scalars. We are also commonly interested in the time derivatives of tensors, and we must similarly demand that these derivatives be themselves tensors.

We have already established the proper form of time derivative of a tensor in terms of the intrinsic derivative, given, for the case of a contravariant vector, by Eq. A.117. In that equation, u_k was taken to be some function of position along a curve given in parametric form as $x_m(t)$. As a special case, we may let the curve represent the path of a particle of material, so that t is time, and

$$\frac{\delta u_k}{\delta t} = u_{k,m} v_m \tag{B.7}$$

If we now allow u_k to be also a function of time at fixed position x_m, then the material derivative of u_k becomes

$$\frac{Du_k}{Dt} = \frac{\partial u_k}{\partial t} + \frac{\delta u_k}{\delta t} = \frac{\partial u_k}{\partial t} + u_{k,m} v_m$$

or

$$\frac{Du_k}{Dt} = \frac{\partial u_k}{\partial t} + v_m \frac{\partial u_k}{\partial x_m} + \left\{ \begin{matrix} k \\ rm \end{matrix} \right\} u_r v_m \tag{B.8}$$

Note that in cartesian coordinates the Christoffel symbols vanish, and Eq. B.8 is identical to Eq. B.6 with f replaced by u_k.

II. DEFORMATION

We shall consider some arbitrary motion existing in a material continuum, and assume that a fixed cartesian coordinate system exists with respect to which we measure position and velocity. We confine our attention to some point **O** at which the velocity components are given by v_{oi}. We take the velocity in the neighborhood of **O** to have components v_i and assume that the motion is continuous, so that v_i approaches v_{oi} as we near the point **O**.

Then we may express the velocity in the neighborhood of **O** by a Taylor series:

$$v_i = v_{oi} + \left(\frac{\partial v_i}{\partial x_j}\right)_o dx_j + \text{terms of magnitude } (dx_j)^2 \tag{B.9}$$

KINEMATICS

If dx_j is a small quantity, we may write

$$v_i - v_{oi} = \left(\frac{\partial v_i}{\partial x_j}\right)_o dx_j \tag{B.10}$$

as the relative velocity components between material at **O** and material in the neighborhood of **O**.

In the Introduction it was stated that we are concerned primarily with internal response, that is, relative motion within the medium. It should be clear that from Eq. B.10 there will be relative motion only if $\partial v_i/\partial x_j$ is not zero. Thus we can take $\partial v_i/\partial x_j$ to be some measure of internal response of the material. We may call it the velocity gradient.

It will be instructive to decompose the velocity gradient into a symmetric and antisymmetric part:

$$\frac{\partial v_i}{\partial x_j} = \frac{1}{2}\left(\frac{\partial v_i}{\partial x_j} + \frac{\partial v_j}{\partial x_i}\right) + \frac{1}{2}\left(\frac{\partial v_i}{\partial x_j} - \frac{\partial v_j}{\partial x_i}\right) \tag{B.11}$$

Let Δ and ω be the symmetric and antisymmetric terms, respectively, so that

$$\Delta_{ij} = \left(\frac{\partial v_i}{\partial x_j} + \frac{\partial v_j}{\partial x_i}\right) \tag{B.12}$$

and

$$\omega_{ij} = \left(\frac{\partial v_i}{\partial x_j} - \frac{\partial v_j}{\partial x_i}\right) \tag{B.13}$$

We wish to examine the nature of the response determined by ω_{ij}. Consider some point **P** such that, relative to the point **O**,

$$P_i = dx_i \tag{B.14}$$

The velocity at **P**, relative to that at **O**, is

$$dv_i = \frac{\partial v_i}{\partial x_j} P_j \tag{B.15}$$

and that part of the motion due to ω is given by

$$(dv_i)_\omega = \tfrac{1}{2}\omega_{ij} P_j \tag{B.16}$$

Now multiply both sides of Eq. B.16 by dx_i, and use Eq. B.14 to obtain

$$(dv_i)_\omega \, dx_i = \tfrac{1}{2}\omega_{ij} \, dx_j \, dx_i \tag{B.17}$$

The indices i and j on the right-hand side of Eq. B.17 are dummy indices and may be reversed, so that

$$\omega_{ij} \, dx_j \, dx_i = \omega_{ji} \, dx_i \, dx_j \tag{B.18}$$

But, by definition, $\omega_{ij} = -\omega_{ji}$, so

$$\omega_{ij}\, dx_j\, dx_i = -\omega_{ij}\, dx_i\, dx_j \tag{B.19}$$

Clearly, this is possible if, and only if

$$\omega_{ij}\, dx_i\, dx_j = 0 \tag{B.20}$$

Now using Eqs. B.16 and B.14, we find

$$(dv_i)_\omega P_i = 0 \tag{B.21}$$

But this is just the condition that two vectors $\mathbf{P}$ and $(\mathbf{dv})_\omega$ be perpendicular. Since $\mathbf{P}$ can be considered an arbitrary position vector, we have shown that that part of the relative motion between two points described by $(\mathbf{dv})_\omega$ is always perpendicular to the line joining the points. But this is just the requirement that the motion be a rigid rotation. We call $\boldsymbol{\omega}$ the *vorticity tensor*.

Rheology is not concerned with rigid rotation, or translation, but rather with *deformation*, the change of relative distance of particles in the material. Hence it must be the symmetric part only of the velocity gradient which describes deformation. We call $\boldsymbol{\Delta}$ the *deformation rate tensor*, and it is calculated, in cartesian coordinates, from the definition in Eq. B.12.

If it is more convenient to work in some curvilinear coordinate system, then we can transform to that system according to the methods outlined in Appendix A. In particular, we note that, in a curvilinear coordinate system, the definition of $\boldsymbol{\Delta}$ must be modified to

$$\Delta^i_j = (v^i_{,j} + v^j_{,i}) \tag{B.22}$$

to retain the correct tensorial character for the derivatives upon transformation.

III. STRAIN

The concept of rate of deformation is inherently simpler than the concept of deformation itself. In discussing rate of deformation there was no need to use any spatial reference state against which to measure kinematic quantities. A deformation rate existed if the distance between two material particles varied with time. On the other hand, a deformation rate may vanish but the material may still be deformed if the material particles do not presently occupy those spatial positions corresponding to a reference configuration defined at an earlier time. It is necessary, then, to look more carefully at the concepts of *configuration*, *motion*, and *reference configuration*.

KINEMATICS

A *configuration* is the name given to the mathematical statement specifying the *places* occupied by material *particles*. It is written

$$\mathbf{x} = \chi(X) \tag{B.23}$$

and it says that, to each particle X of a body there corresponds a certain place, in a fixed coordinate system $\mathbf{x}$. If the function χ is changed, then the configuration has changed. If χ varies with time, then motion occurs as the body's configuration changes, and

$$\mathbf{x} = \chi(X, t) \tag{B.24}$$

defines a *motion*. The notation is changed from that of Eq. B.2 to emphasize the importance of *configuration* defined by the function χ.

If some particular configuration statement, say $\chi = \varkappa$, is given as a *reference* configuration, then

$$\mathbf{x} = \varkappa(X) \tag{B.25}$$

defines the places occupied by material particles in the reference configuration. Once a reference configuration is specified, a set of material coordinates $\mathbf{X}$ may be defined as the places $\mathbf{x}$ occupied by the material particles in the reference configuration, or

$$\mathbf{x} = \mathbf{X} = \varkappa(X) \tag{B.26}$$

Once a reference configuration is defined it is possible to express a motion as

$$\mathbf{x} = \chi(\mathbf{X}, t) \tag{B.27}$$

where it is understood that $\mathbf{X}$ is given in the reference configuration $\varkappa$.

It is possible to specify a varying reference configuration, rather than some fixed configuration. For example, the reference configuration at time t may be the configuration itself at *that time*, or

$$\mathbf{x} = \chi(X, t)$$

Then, at another time t', the same particles are in the places

$$\boldsymbol{\xi} = \chi(X, t') \tag{B.28}$$

$\boldsymbol{\xi}$ is the place, at time t', occupied by a particle which was at the place $\mathbf{x}$ at time t. An alternate notation is

$$\boldsymbol{\xi} = \chi_{(t)}(\mathbf{x}, t') \tag{B.29}$$

A region of a material will be said to have undergone a *local strain* if two neighboring material particles suffer a change of separation. In the fixed reference configuration **X** the separation (squared) is given, in terms of cartesian components, by

$$(dS)^2 = dX_k \, dX_m \, \delta_{km} \tag{B.30}$$

At some later time the particles would be separated by a (squared) distance

$$(ds)^2 = dx_i \, dx_j \, \delta_{ij} \tag{B.31}$$

where the coordinates **x** are the places occupied by material particles after the motion $\chi(\mathbf{X}, t)$. Since **x** and **X** are related through the motion, the chain rule for differentiation may be used to show that

$$(ds)^2 = \frac{\partial x_i}{\partial X_k} \frac{\partial x_j}{\partial X_m} \delta_{ij} \, dX_k \, dX_m \tag{B.32}$$

Then the change in separation is given by

$$(ds)^2 - (dS)^2 = \left(\frac{\partial x_i}{\partial X_k} \frac{\partial x_j}{\partial X_m} \delta_{ij} - \delta_{km} \right) dX_k \, dX_m \tag{B.33}$$

If a tensor **E** is defined, with cartesian components

$$E_{km} = \frac{\partial x_i}{\partial X_k} \frac{\partial x_j}{\partial X_m} \delta_{ij} \tag{B.34}$$

then inspection of Eq. B.33 reveals that a condition of vanishing strain requires that $\mathbf{E} = \boldsymbol{\delta}$. The tensor **E** is a strain tensor.

If a similar analysis is carried out for a deformation referred to the *present* configuration, it will be found that a suitable definition of a strain tensor would be

$$C_{ij}^{(t)} = \frac{\partial \xi_k(t')}{\partial x_i(t)} \frac{\partial \xi_m(t')}{\partial x_j(t)} \delta_{km} \tag{B.35}$$

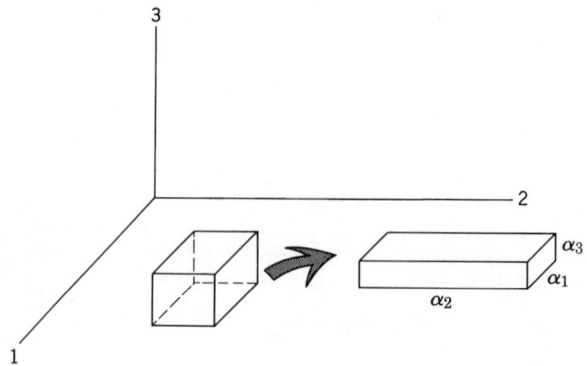

Figure B.1

KINEMATICS

$\mathbf{C}^{(t)}$ is usually called the *relative* Cauchy-Green strain tensor, and its non-cartesian components are defined in Eq. 3.33. The adjective "relative" and the superscript (t) have been dropped, in most of the text, with the addition of a statement that the *present* configuration is the reference configuration.

As an example consider a deformation defined, with respect to the configuration at time t, by

$$
\begin{aligned}
\xi_1(t') &= \alpha_1 x_1(t) \\
\xi_2(t') &= \alpha_2 x_2(t) \qquad \text{for} \quad t' < t \\
\xi_3(t') &= \alpha_3 x_3(t)
\end{aligned}
$$

$$
\begin{aligned}
\xi_1(t') &= x_1(t) \\
\xi_2(t') &= x_2(t) \qquad \text{for} \quad t' \geq t \\
\xi_3(t') &= x_3(t)
\end{aligned}
\tag{B.36}
$$

This is an elongational deformation in which the material is at rest for some period of time, and then is suddenly stretched to a new rest configuration. The constants α_1, α_2, and α_3 are the stretch ratios, or principal extension ratios. Figure B.1 shows this deformation.

If the current configuration is taken for reference, the Cauchy-Green tensor is (using Eq. B.35)

$$
\mathbf{C} = \begin{pmatrix} \alpha_1^2 & 0 & 0 \\ 0 & \alpha_2^2 & 0 \\ 0 & 0 & \alpha_3^2 \end{pmatrix} \qquad \text{for} \quad t' < t
$$

$$
\mathbf{C} = \boldsymbol{\delta} \qquad \text{for} \quad t' \geq t
$$

The result $\mathbf{C} = \boldsymbol{\delta}$ is simply a reflection of the fact that the present configuration is taken as the reference configuration.

Formal development of kinematical concepts may be found in the following books:

B1. Prager, W., *Introduction to the Mechanics of Continua*, Ginn, Boston, 1961, Ch. 9.
B2. Truesdell, C., and W. Noll, "The Non-linear Field Theories of Mechanics," in *Handbuch der Physik*, vol. III/3, S. Flügge, Ed., Springer, Berlin, 1965, pp. 48–52.

Appendix C

Monomer Structural Units of Some Common Polymers

1. Polyacrylic acid

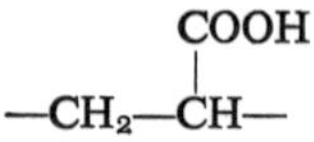

$$-CH_2-\underset{\underset{COOH}{|}}{CH}-$$

2. Polyacrylonitrile

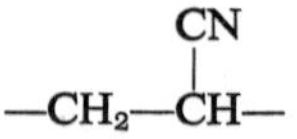

$$-CH_2-\underset{\underset{CN}{|}}{CH}-$$

3. Polybutadiene

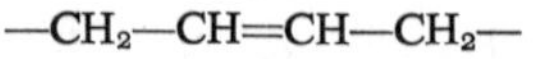

$$-CH_2-CH=CH-CH_2- \qquad -CH_2-\underset{\underset{CH=CH_2}{|}}{CH}-$$
$$(1,4) \qquad\qquad (1,2)$$

4. Polyisobutylene

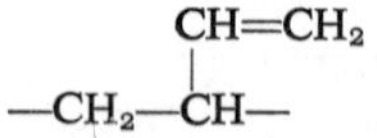

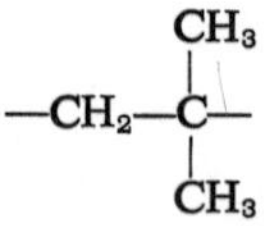

$$-CH_2-\underset{\underset{CH_3}{|}}{\overset{\overset{CH_3}{|}}{C}}-$$

5. Poly-1-butene

$$-CH_2-\underset{\underset{C_2H_5}{|}}{CH}-$$

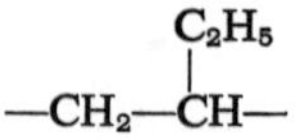

6. Polydimethylsiloxane

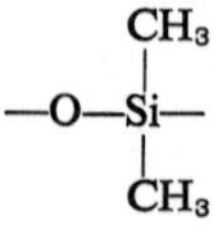

$$-O-\underset{\underset{CH_3}{|}}{\overset{\overset{CH_3}{|}}{Si}}-$$

7. Polyethylene oxide

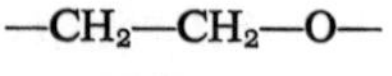

$$-CH_2-CH_2-O-$$

8. Polyethylene

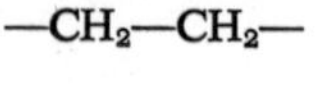

$$-CH_2-CH_2-$$

9. Polyisoprene

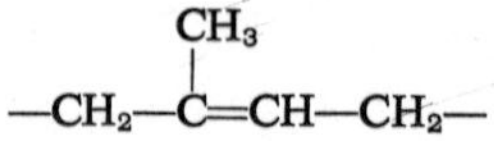

$$-CH_2-\underset{\underset{}{}}{\overset{\overset{CH_3}{|}}{C}}=CH-CH_2- \qquad -CH_2-\underset{\underset{CH=CH_2}{|}}{\overset{\overset{CH_3}{|}}{C}}-$$

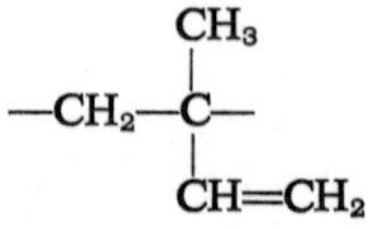

MONOMER STRUCTURAL UNITS

10. Polymethyl methacrylate

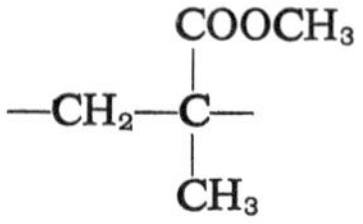

$$-CH_2-\underset{\underset{CH_3}{|}}{\overset{\overset{COOCH_3}{|}}{C}}-$$

11. Polystyrene

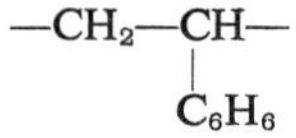

$$-CH_2-\underset{\underset{C_6H_6}{|}}{CH}-$$

12. Polyvinylacetate

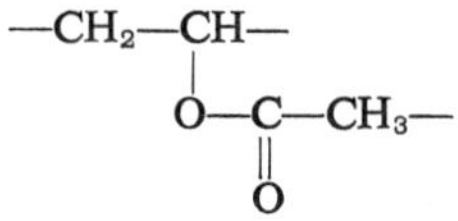

$$-CH_2-\underset{\underset{O-\underset{\underset{O}{||}}{C}-CH_3-}{|}}{CH}-$$

13. Polyvinylchloride

$$-CH_2-\underset{\underset{Cl}{|}}{CH}-$$

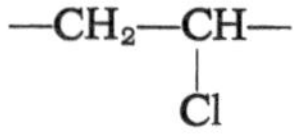

14. Polypropylene

$$-CH_2-\underset{\underset{CH_3}{|}}{CH}-$$

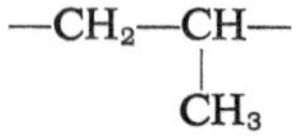

AUTHOR INDEX

Numbers in parentheses are reference numbers and show that an author's work is referred to although his name is not mentioned in the text. Numbers in *italics* indicate the pages on which the full references appear.

Abramowitz, M., 109(45), 111(48), 112 (45,48), *131*
Acrivos, A., 100(16), *130*
Ajroldi, G., 194, 195(50), *199*
Alfrey, T., 110, *131*
Allen, V. R., 168(1), 170(1,5), 171, *198*
Aloisio, C. J., 195(54), 197(54), *199*
Andrews, R., 111(49), *131*
Appeldorn, J. K., 30(12), *82*
Aris, R., *229*
Ashare, E., 192(49), *199*
Astarita, G., 49, 50, *82*

Ballman, R., 64–66, *83*, 134(4), 155, *167*, 179, *198*
Barnett, S., 100(20), *130*
Baronet, C. N., 52(28), *83*
Bauer, J. W., 175(17), 180(17), *198*
Bauer, W. H., 149
Bellinger, J. C., 77, *83*
Berge, J. W., 76, *83*
Bernstein, B., 120, 128, *131*
Billmeyer, F. W., 134, 144(2), *166*
Bird, R. B., 7, *12*, 14(2), 15(2), 30, 32(8), 35, 36(13), 37, 71(47), *82*, *83*, 99(12), 100(19,22), 102, 117, *130*, *131*, 176(18), 192, 193(48), *198*, *199*
Bogue, D. C., 99, 117, 120, *130*, 186(33), *198*
Boltzmann, L., 115, *131*
Boyce, R. J., 175(17), 180(17), *198*
Boyer, R. F., 171, *198*
Brandrup, J., 145(13), *167*
Brinkman, H. C., 30, *82*
Brodnyan, J. G., 101(29), 129(29), *130*, 153(22,23), *167*, 182, *198*
Brown, D. R., 60(36), *83*, 185(36), 186, 187(36), *199*
Browning, G., 172(12), *198*

Bueche, A., 142–144, 161, *167*
Bueche, F., 147, 148, *167*, 170, *198*
Busse, W., 172, 174(14), 183, *198*

Cantow, M. J. R., 154, 155(24), *167*
Catsiff, E., 77, *83*, 113, *131*
Chao, J.-L., 49(26), *82*
Chinai, S., 172(13), *198*
Christopher, R., 100(17), *130*
Coleman, B. D., 8(3), *12*, 22(5), 60(37), *82*, *83*, 94(6), 96, 97, *130*
Collins, E., 149, 175(17), 180(17), *198*
Colwell, R. E., 13(1), *82*
Cox, W. P., 70, *83*
Craft, T., 120–122, *131*

Dahler, J. S., 5(1), *12*
Debye, P., 140–144, *167*
DeHoff, P. H., 123, *131*
DeVries, A. J., 194, *199*
DeWitt, T. W., 57, *83*, 105(42), *131*, 151, 181(29), 190(29), *198*
Dillon, R., 102(33), *131*
Dodge, D. W., 100(14), *130*
Doughty, J. O., 99(11), 117, *130*
Dunell, B., 111(49), *131*
Dunleavy, J., 151(21), *167*, 172(10), 181(10), 182(10), *198*

Einstein, A., 139, *167*
Eisner, Y., 163, *167*
Elrod, H., 22(6), *82*
Elyash, L., 151(20), *167*, 179, *198*
Ende, F., 69 (44), *83*
Ericksen, J. L., 94, *130*
Eringen, A. C., 91(3), 125(3), *130*
Evans, W., 171(8), 192(48), 193(48), *198*, *199*
Eyring, H., 101(25–27), *130*

Ferry, J. D., 67(42), 71, *83*, 111, *131*, 163, 164, 165(43), 166(43), *167*, 171, 172, 176, 178(19), 190(44–46), 191(44), 192(48), 193(48), *198*, *199*
Finger, J., 128, *131*
Fixman, M., 159, 163(41), *167*
Flory, P., 134, 135(1), 136(1), 147(1), *166*
Foster, E., 172(12), *198*
Fox, T. G, 168(1), 170, 171, 178, *198*
Frazer, W., 151(20), *167*, 179, *198*
Frederick, J. E., 164, 165(43), 166(43), *167*, 190(46), *199*
Fredrickson, A. G., 18(4), *82*, 99(9), 116, 119, *130*
Frisch, H. L., 139, *167*
Fujii, T., 105(41), *131*, 189(40), *199*
Fujita, H., 174(16), *198*

Garbuglio, C., 194, 195(50), *199*
Gaskins, F., 101(29), 129(29), *129*, 150
Gavis, J., 44(19,20), 46, 78(56,57), 79, 81, *82*, *83*
Gibbs, D. A., 75, *83*
Giesekus, H., 52(29), *83*
Gill, S. J., 78(56), *83*
Gillespie, T., 102(32), *131*
Ginn, R. F., 47(24), 56, 57(24), 81(24), *82*, 185(34), 186, 187(34), *198*
Gogos, C., 196, *199*
Goldberg, W., 123(57), *131*
Goren, S., 79, 81, *83*
Graessley, H. H., 157, 158, *167*
Grandine, L., 172(11), *198*
Gratch, S., 178(20), *198*
Greensmith, H. W., 54(31), 57(31), *83*
Griffith, R. M., 100(18), *130*
Gruver, T., 179, *198*

Hallman, T. M., 30(10), *82*
Harrison, G., 162(37), *167*, 190, 191(41), *199*
Hayes, J. W., 50, *82*
Hazelton, R. L., 158(28), *167*
Hermans, J. J., 143, *167*
Hildebrand, F. B., 15(3), *82*
Hoftyzer, P., 145(14), *167*
Holde, K. van, 74, *83*
Holden, G., 180(24), *198*

Holmes, L. A., 190(44), 191(44), 192(49), *199*
Hoppmann, W. H., II, 52(28), *83*
Humphrey, A., 100(20), *130*
Huppler, J. D., 47(23), 50, *82*, 117, 119, *131*, 160(34), *167*, 192(48,49), 193(48), *199*

Immergut, E. H., 145(13), *167*
Ito, Y., 181, *198*

Jahnke, E., 69(44), *83*
Jeffreys, H., *229*
Johnson, G. C., 149
Johnson, J. F., 151(19), 153(19), 154, 155(24), *167*, 170(2), 174(16), *198*
Johnson, M., 171(8), *198*
Jones, R. W., 76, *83*
Jordan, I., 171(8), *198*

Karam, H. J., 77, *83*
Kataoka, T., 34(12a), *82*
Kato, H., 105(41), *131*, 189(40), *199*
Katsuhisa, T., 185(39), 189(39), 190(39), *199*
Kearsley, E., 120, 128, *131*
Keeney, R., 70, *83*
Kelley, E. L., 153(22,23), *167*, 182, *198*
Kim, K. Y., 13(1), *82*
Kimura, S., 171(9), *198*
Kirkwood, J., 142, 143, 160, *167*
Kotaka, T., 55(32), *83*, 185(37,42), 186, 188(38), 189(38), 190(38,43), *199*
Kraus, G., 179, *198*
Krieger, I. M., 22(6), *82*
Kurata, M., 55(32), *83*, 185(32,37,39,42), 186(37), 188(38), 189(38,39), 190(32, 38,39,43), *198*, *199*

Lamb, J., 162(37), *167*, 190(41), 191(41), *199*
Landel, R., 176, 178(19), *198*
LaNieve, H., 186(33), *198*
Leaderman, H., 73(48), 76, *83*, 177, 178, *199*
Lianis, G., 123(57), *131*
Lightfoot, E. N., 7(2), *12*, 14(2), 15(2), *82*, 176(18), *198*

AUTHOR INDEX

Lindeman, L. R., 158(28), *167*
Lindgern, E. R., 49(26), *82*
Litt, M., 100(20), *130*
Longworth, R., 172, 174(14), 183, *198*
Losch, F., 69(44), *83*
Loshaek, S., 178, *198*
Lyons, J. W., 13(1), *82*

McConnell, A. J., *229*
Macdonald, I. F., 192(49), *199*
McEachern, D. W., 100(21), *130*
McKelvey, J. M., 181, *198*
Malkin, A. Ya., 194, *199*
Markovitz, H., 56(33), 59, 60, 68(43), *83*, 105(42), *131*, 181(29), 185(36), 186, 187(36), 190(29), *198*, *199*
Masuda, T., 171(9), *198*
Matheson, A., 162(37), *167*, 190(41), 191(41), *199*
Matsuoka, S., 195(54), 197(54), *199*
Maxwell, B., 195(54), 197(54), *199*
Maxwell, J. C., 102, *131*
Mendelson, R., 134(3), *166*, 181, *198*
Merrill, E. W., 75, *83*, 101(28), *130*
Meter, D. M., 102(31), *131*
Metzner, A. B., 43(18), 47, 49, 50, 53, 54(18), 56, 57(24), 81(24), *82*, 100(14,15), *130*, 185(34), 186, 187(34), *198*
Mickley, H. S., 101(28), *130*
Middleman, S., 38(15), 40(15), 44(19,20), 78(57), 79(57,59), 81(59), *82*, *83*, 100(17), *130*, 150(18), 151(21), 153(18), 154(18), 156(18), *167*, 172(10), 181(10), 182(10), *198*
Miller, C. E., 52(28), *83*
Miyanaga, N., 171(9), *198*
Modan, M., 46, *82*
Morawetz, H., 145(15), 146(15), *167*
Murakami, K., 108, *131*

Nielsen, L. E., 70, *83*
Ninomiya, K., 190(44), 191(44), *199*
Noll, W., 8(3), *12*, 22(5), 41(16), 60(37), *82*, *83*, 91, 94, 96, 97, *130*, *237*

Ogihara, S., 105(41), *131*, 189(40), *199*
Oka, S., 67(40), *83*
Oldroyd, J. G., 85, 88, 104, 116, *130*, *131*

Onogi, S., 105(41), *131*, 171(9), 189(40), *198*, *199*
Osaki, K., 185(32,39), 188(38), 189(38,39), 190(32,38,39,43), *198*, *199*
Otto, R., 100(15), *130*

Padden, F., 57, *83*, 105(42), *131*, 181(29), 190(29), *198*
Petersen, E. E., 100(16), *130*
Peticolas, W. L., 196, *199*
Pezzin, G., 194, 195(50), *199*
Philippoff, W., 71(46), *83*, 101(29), 129(29), *130*, 150
Plazek, D. J., 76, *83*
Porter, R. S., 151(19), 153(19), 154, 155(24), *167*, 170(2), 174(16), *198*
Powell, R. E., 101(27), *130*
Powell, R. L., 38(15), 40(15), 48(21), 81(21), *82*
Prager, W., 91(4), 125(4), *130*, *237*
Prandtl, L., 101(24), *130*
Prozorovskaya, N., 180, 186(25), *198*
Ptitsyn, O., 163, *167*

Ram, A., 101(28), *130*
Rautenbach, E., 79(59), 81(59), *83*
Ree, F. H., 101(25), *130*
Ree, T., 101(25,26), *130*
Reed, J. C., 100(14), *130*
Riseman, J., 142, 143, 160, *167*
Rivlin, R. S., 54(31), 57(31), *83*, 94, *130*
Robinson, D., 155, 156, *167*
Rouse, P. E., 104(39), *131*, 161, *167*

Sadowski, T. J., 100(22,23), *130*
Sadron, C., 139, *167*
Savins, J. C., 48, 53, *82*
Sawyer, W., 172(12), *198*
Schneider, W., 172(13), *198*
Schremp, F. W., 192, 193, *199*
Scriven, L. E., 5(1), *12*
Seely, G., 102(34), *131*, 181, 183(30), *198*
Shah, M. J., 100(16), *130*
Shertzer, C. R., 43(18), 47, 53, 54(18), *82*, 186, *198*
Siegel, R., 30, *82*
Simha, R., 139, *167*, 171, *198*
Simon, R., 134(4), 155, *167*, 179, *198*

Slattery, J. C., 25(7), *82*, 99(12), *130*
Smith, K. A., 49(26), *82*
Smith, R. G., 76, *83*
Smith, T. L., 62, *83*
Spain, B., *229*
Sparrow, E. M., 30(10), *82*
Spencer, R. S., 102(33), *131*
Spriggs, T. W., 104, 105(36), 106, 117, *131*, 192(48,49), 193, *199*
Steidler, F. E., 30(12), *82*
Stewart, W. E., 7(2), *12*, 14(2), 15(2), *82*, 176(18), *198*
Stockmayer, W. H., 101(28), *130*, 163(41), *167*
Stratton, R. A., 155, 156, *167*
Sutterby, J. L., 99(13), 102(13,30), *130*

Takashi, K., 171(9), *198*
Tamura, M., 55(32), *83*, 185, 186(37), 188(38), 189(38,39), 190(32,38,39,43), *198*, *199*
Tanaka, K., 185(32), 190(32), *198*
Tanner, R. I., 50, *82*
Temple, G., *229*
Teramoto, A., 174(16), *198*
Thomas, D., 140, *167*
Tobolsky, A. V., 77, *83*, 108, 111(49), 112, 113, *131*
Tochon, J., 194, *199*
Toor, H. L., 30, *82*
Trouton, F. T., 11, *12*
Truesdell, C. A., 41(16), *82*, 85, *130*, *237*
Tschoegl, N. W., 163, 164, 165(43), 166(43), *167*, 190(45,46), *199*
Turian, R. M., 35, 36(13), 37, *82*

Udy, D., 172(11), *198*
Ueda, S., 34(12a), *82*

Van Krevelen, D., 145(14), *167*
Van Wazer, J. R., 13, *82*
Vinogradov, G., 180, 186, 194, *198*, *199*
Vrancken, M. N., 76, *83*

Williams, J. W., 74, *83*
Williams, M. C., 42, 51, 53, 71(47), *82*, *83*, 152(30), 158–160, *167*
Williams, M. L., 111, *131*, 172, 176, 178(19), 181, *198*
Wylie, C. R., Jr., 67(41), *83*
Wyman, D., 151(20), *167*, 179, *198*

Zapas, L., 105(42), 120–122, 128, 129, *131*, 181(29), 190(29), *198*
Zimm, B. H., 104(40), *131*, 160–162, *167*

SUBJECT INDEX

Apparent viscosity, 10
Axial annular flow, 50
Axial thrust, jet, 42
   cone and plate, 53
   parallel plate torsion, 55

Barus phenomenon, 41
BKZ theory, 120
Boltzmann superposition principle, 115
Branching, 133
Bueche theory, 147

Capillary jet, 42
Capillary viscometer, 13
   entrance effects, 15
   viscous heating, 29
Cauchy-Green tensor, 92, 237
Cayley-Hamilton theorem, 211
Christoffel symbols, 220
Coaxial cylinder viscometer, 19
Complex compliance, 71
Complex shear modulus, 70
Complex viscosity, 67
Cone and plate flow, inertial effects, 51
   normal stresses, 50
   secondary flows, 52
   viscosity, 25
   viscous heating, 37
Configuration, 235
Continuity equation, 3
Contravariant components, 217
Convected coordinates, 85
Coordinate transformation, 215
Couette flow, 19
   inertial stresses, 59
   normal stresses, 57
   viscous heating, 35
Covariant components, 217
Covariant derivative, 223
Creep, 71
Creep compliance, 73
Curvilinear base vectors, 215

Degree of polymerization, 133

Die swell, 41
Displacement gradient, 91
Dissipation, 7
Distribution, of molecular weight, 133
   of relaxation times, 109
Dynamic equations, 3
Dynamic measurements, 65
Dynamic rigidity, 70
Dynamic viscosity, 67, 160

Elastic potential, 125
Ellis fluid, 100
Elongation, 60, 63
Elongational viscosity, 10, 11, 64
Energy equation, 7, 29
Entrance length, normal stresses, 45
   shear stresses, 15
Excluded volume, 135, 163

Finger strain tensor, 117
First-order fluid, 97
Free draining coil, 140

Glass temperature, 171
Graessley's theory, 156

Integral theories, 114, 117
Intrinsic viscosity, 142
Invariants, 213

Jaumann derivative, 91
Jet expansion, 43
   effect of viscous heating on, 46

Loss modulus, 71, 188

Material coordinates, 231
Material derivative, 231
Maxwell fluid, 102, 106, 164
Mean normal stress, 42
Memory function, 98
Metric tensor, 216
Molecular weight distribution, 133
Monomer friction coefficient, 141

SUBJECT INDEX

Motion, 230

Normal stress coefficients, 10, 184
   dependence on shear rate, 186
Normal stress measurement, 40
   capillary flow, 42
   cone and plate flow, 50
   parallel plate torsion, 54
   pitot tube, 48
Number average molecular weight, 133

Oldroyd derivative, 88
Oscillatory flow, 67

Parallel plate torsion, 54
   inertial effects, 55
Parallel plate viscoelastomer, 74
Physical components, 226
Pitot tube measurement, 48
Poiseuille flow, 13
Polydispersity, 146
   effect on viscosity, 150
Powell-Eyring fluid, 101
Power law fluid, 19, 100
   viscous dissipation, 30
Prandtl-Eyring fluid, 101
Principal axes, 210
Purely viscous fluid, 99

Rabinowitsch-Mooney equation, 15
Rate of deformation tensor, 6
Rate equations, 102
Reduced viscosity, 148
Relaxation modulus, 76
Relaxation time, Bueche theory, 148
Retarded elasticity function, 73
Rivlin-Ericksen fluid, 94
Rouse theory, 108, 161

Second-rank tensor, 205
Shear rate, 9
Shear relaxation modulus, 113
Shielding parameter, 142
Shift factor, between steady and
     dynamic viscosity, 105, 188
   for relaxation time, 149, 176
Simple elongation, 10, 60
Simple fluid, 91

Simple shear flow, 8
   notation, 9
   of a simple fluid, 92
Spatial coordinates, 231
Sprigg's model, 104
Staudinger's viscosity rule, 142
Steady shear compliance, 73
Storage modulus, 71, 187, 190
Strain energy function, 125
Strain rate, 10
Strain tensor, 86, 236
   Cauchy-Green, 92, 237
   Finger, 117
Stress relaxation, 192
   upon cessation of steady flow, 192
   effect of molecular weight distribution, 196

Tensile creep, 77
Tensile viscosity, 11
Theta solvent, 145
Torsional pendulum, 76
Trouton viscosity, 11

Upper newtonian viscosity, 101, 129

Viscoelastic fluid, 102
Viscometer, capillary, 13
   cone and plate, 25
   Couette, 19
Viscometric flows, 8
Viscosity, concentration dependence, 171
   molecular weight dependence 169
   shear rate dependence, 178
   temperature dependence, 175
Viscosity average molecular weight, 146
Viscous dissipation, 28
   approach to equilibrium, 38

Wave propagation, 77
Weissenberg effect, 41
Weissenberg-Rabinowitsch-Mooney
     equation, 15
Weissenberg relation, 56, 105
Williams' theory, 158
WLF equation, 176

Zimm's theory, 160